SATISFYING REASON

Episteme

A SERIES IN THE FOUNDATIONAL, METHODOLOGICAL, PHILOSOPHICAL, PSYCHOLOGICAL, SOCIOLOGICAL, AND POLITICAL ASPECTS OF THE SCIENCES, PURE AND APPLIED

VOLUME 21

The titles published in this series are listed at the end of this volume.

NICOLAS RESCHER
University of Pittsburgh

SATISFYING REASON

Studies in the Theory of Knowledge

KLUWER ACADEMIC PUBLISHERS
DORDRECHT / BOSTON / LONDON

Library of Congress Cataloging-in-Publication Data

Rescher, Nicholas.
Satisfying reason : studies in the theory of knowledge / by Nicholas Rescher.
p. cm. -- (Episteme ; v. 21)
Includes index.
ISBN 0-7923-3148-6 (alk. paper)
1. Rationalism. I. Title. II. Series: Episteme (Dordrecht, Netherlands) ; v. 21.
BD181.R45 1995
121--dc20 94-33491

Published by Kluwer Academic Publishers,
P.O. Box 17, 3300 AA Dordrecht, The Netherlands.

Kluwer Academic Publishers incorporates
the publishing programmes of
D. Reidel, Martinus Nijhoff, Dr W. Junk and MTP Press.

Sold and distributed in the U.S.A. and Canada
by Kluwer Academic Publishers,
101 Philip Drive, Norwell, MA 02061, U.S.A.

In all other countries, sold and distributed
by Kluwer Academic Publishers Group,
P.O. Box 322, 3300 AH Dordrecht, The Netherlands.

Printed on acid-free paper

Printed in the Netherlands

TABLE OF CONTENTS

PREFACE

The dozen essays on issues in the philosophy of knowledge that are collected together were written during the 1991–93 triennium. Two of them (numbers 2 and 8) have previously appeared in print in professional journals and I am grateful to the publishers involved for permission to reprint. I am also grateful to Marian Kowatch and Annamarie Morrow for preparing a typescript of this material serviceable for the printer's needs.

May 1993

Nicholas Rescher
Pittsburgh, PA

INTRODUCTION

Leibniz said with a mixture of admiration and exasperation that the duchess Sophie of Hannover always wanted to know the reason why behind the reason why. And that is just how rationality works: it wants to leave no loose ends to understanding, seeking to enable us to understand things through to the bitter end.

The present book presents a dozen essays dealing with issues of cognitive rationality from one perspective or another. The opening essay ("Satisfying Reason") considers the changes of perspective that have, over the centuries, led philosophers to soften the requirements of reason through a step-by-step weakening of the absolutistic and necessitarian demands of the great thinkers of the 1600s. This softening has involved a fundamental change in our understanding of what cognitive rationality is all about.

The second essay ("Why Be Rational?") shows how cogent and compelling the case for heeding the demands of reason actually are, once they themselves are reasonably understood. There is no *reasonable* way of dismissing the demands of reason.

The third essay ("Reason and Reality") argues that despite the compelling case for cognitive rationality one must face the harsh realities of our epistemic situation. The project of rational inquiry is much like that of morality. We humans are committed in both regards to pursue what is at best and at most a life of effort and struggle that offers the prospect of achieving not perfection but merely ongoing improvements.

Insofar as metaknowledge–knowledge about our knowledge–is achievable at all (so argue the next two essays), we must see ourselves as striving in scientific inquiry to achieve best estimates whose ultimate correctness is nowise assured. Our condition is such that in the cognitive life there are no guarantees with regard to the big theoretical issues to which science addresses itself. We have to do the best we can, realizing that it may ultimately transpire that our best has not been good enough.

What is called for in this situation (so argues the sixth essay on "Methodological Optimism") is a positive and constructive attitude that grants a presumption of adequacy to our established procedures. And what recommends this methodological optimism is exactly its positivity: the fact that if good results are to be achievable by us at all, then this way of proceeding is–on all available indications–the most effective way to realize them.

The remaining six chapters address various issues relating to the use and abuse of reason.

Chapter seven criticizes the misguided tendency to think that the domain of clear and adequate understanding is confined to those issues that can be treated quantitatively. Measurement is all very well in its place, but this is not ubiquitous throughout the domain of rigorous thought.

Chapter eight argues that a proper understanding of the world we live in calls for endeavoring to grasp it not in terms of abstract necessities but rather in terms of familiar, first-hand accessible categories of the human mind and its experiential repertoire.

The next chapter develops further the contrast between causal/explanatory understanding and experiential/normative understanding. It stresses the idea that *Homo sapiens* as a member of both the *rational* and also the specifically *human* order of things, must adopt a view of the world and our place in it that takes account of both the scientifically explanatory and the humanistically affective dimensions.

The tenth chapter depicts "the definitive truth" about reality as an idealization that contrasts with our putative truths–our *estimates* of the real truth of things as we are best able to develop them.

Since that this sort of estimation is a matter of the informational context, the issue becomes colored with a tinge of relativism. But this contextuality certainly does not warrant an *indifferentist relativism* of arbitrary choice. Insofar as we not only form beliefs but act on them, the situation is that of a functuationalism where we may make our own proposals but the actual disposing lies elsewhere–with the world's humanity-indifferent *modus operandi*. Rational inquiry is certainly no matter of free construction as we have it in belles lettres.

Chapter eleven maintains that indifferentist relativism is an inappropriate cognitive stance even in the interpretation of texts. Texts (literary and other) are not infinitely plastic materials with which we interpreters are free to do as we will. Texts have *contexts* which every sensible interpreter must take into appropriate account.

The final chapter seeks to came to grips with the paradoxical character of some of what we putatively know that is inherent in the fact that our convictions regarding matters of truth and reality are often mutually incompatible. The orthodox strategy here is to insist that when this happens the culprit contentions involved are "semantically meaningless" by way of lacking a truth value. The present discussion argues that this approach is not optimal. The alternative of *accepting* inconsistency and shifting from truth-value *gaps* to truth-value *overlaps* is not only feasible but enjoys certain theoretical advantages. The inconsistency of our putative knowledge is not an absolute disaster; it is something that even fully rational people can take in stride and learn to live with.

Throughout, the book seeks to develop and defend the following perspective:

- that rationality is a cardinal virtue in cognitive matters;
- that this is not something simple and cut-and-dry. In the pursuit of truth through the development of knowledge we face obstacles–sometimes even insuperable ones;

- all that we can do, here as elsewhere, is to do the best we can, realizing that even our very best may still be imperfect;
- that nevertheless the venture is far from hopeless: while absolutes are unattainable in the cognitive venture, some solutions are situationally optimal in being the comparatively best that can be managed in the circumstances;
- that reason itself enables us to come to terms with this state of affairs and urges us to accept the best that we can do as good enough.

Idealized rationality, like idealized morality, may require more of us than we can ever effectively accomplish. But rationality is also realistic. In urging us to see things as they are, reason is prepared to face up to its own limitations. Insofar as we seek to meet the demands of reason, reason itself insists upon our resting satisfied with the subideal best that can practicably be achieved in the prevailing circumstances.

CHAPTER 1

SATISFYING REASON

Some Revolutionary Changes in the Principle of Sufficient Reason

1. REASON'S DEMAND FOR UNDERSTANDING

Homo sapiens qualifies as a rational animal not only because we humans possess the capacity for reason, but also because we are generally reason-demanding, looking for explanations for everything, small or large, that goes on about us. And the impetus to understanding – the quest for an explanation of the world's facts – lies at the core of our rationality. Confronted by any significant fact, our mind, by its very nature, seeks to know "the reason why." It stands committed to the idea encapsulated in Leibniz's *principium rationis sufficientis* – the "Principle of Sufficient Reason" to the effect that there is always explanation for why things are as they are – a basis for understanding why it is that what is so is so and what is not is not. As Leibniz puts it in sec. 32 of the *Monadology*:

> [A principle] of sufficient reason [obtains], in virtue of which we consider that no fact could be true or actual, and no proposition true, without there being a sufficient reason for its being so and not otherwise, although most often these reasons cannot be known by us.

The principle is tantamount to the thesis that for every fact there exists a cogent explanation for its being exactly as it is.

There is much plausibility to this conception. And ever since the dawn of philosophy there have been "rationalists" – people who stand committed to the idea that our "why?" questions must always have a satisfactory answer in principle – even though we may not be able to find it in practice owing to limitations in our cognitive resources. This idealistic theme of the explicability of the real – the inherent intelligibility of things – plays a prominent role in the history of philosophy – from antiquity onwards. Central to the philosophy of Leibniz, it also became a centerpiece of Hegel's philosophy.[1] In the present century its most enthusiastic exponent has been Brand Blanshard.[2] But while the principle has been maintained throughout, it has undergone a process of constant reappraisal and reformulation. The present discussion will consider some of the most radical transformations that have, over time, characterized our philosophical understanding of what it takes to satisfy reason.

To begin with, it deserves note that the Principle of Sufficient Reason is alive and well – that even now we have not altogether given up on it. A visceral sense of this may be conveyed by an illustration. Suppose there is a plane crash, to give an only too familiar example. An investigating team is sent out to inquire into the disaster. In anticipating its ultimate report, we are prepared for a response ranging across a wide spectrum: mechanical failure, sabotage (by a disgruntled employee), bombing (by a terrorist of some description), pilot error, maintenance inadequacies, support-personnel deficiencies (on one cross-Mediterranean flight in the 1960s, El Al's ground personnel neglected to put fuel in the plane). The range of possible explanations is wide and varied. We would even be steeled for an epistemically negative report to the effect that investigators have in the circumstances found themselves unable to identify the cause of failure. But we are *not* prepared for a report that says: "There just is no answer to the explanatory question. This simply happens to be one of those crashes that has no explanation at all." We are simply not willing to accept a distinction like:

- crashes whose occurrence has some sort of explanation;

- crashes that are inherently inexplicable – that have to be attributed (at most and best) to "the general cussedness of things."

This line of reflection clearly indicates that the Principle of Sufficient Reason continues to exert its appeal. But its status and character have changed a great deal over the years, and it is this aspect of the matter that mainly concerns us here.

2. Theological Totalitarianism

The medieval Neo-Platonic/Christian stance towards explanation proposed to trace all factual truth back to a unique first cause. As it viewed the matter, there is, in the final analysis, one single fact – namely, God's creation choice – that provides the ultimate explanation of every aspect of the world's nature. This single fact provides for the rational explanation of everything else.

In just this way the Principle of Sufficient Reason was, for Leibniz, in effect a principle of *optimality*, asserting that the real is exactly that one among the possible alternatives whose reasons for being the case outweigh the reasons against. God chooses to actualize the best of possible worlds: the sufficient reason for all factual truth resides in a Principle of Perfection that codifies God's commitment to optimizing the creation. As he himself puts it:

> All contingent propositions have sufficient reasons, or equivalently have a priori proofs which establish their certainty, and which show that the connection of subject and predicate of these propositions has its foundation in their nature. But it is not the case that contingent propositions have demonstrations of necessity, since their sufficient reasons are based on the principle of contingence or existence, i.e., on what seems best among the equally possible alternatives. (*Discourse on Metaphysics*, sec. 13.)

What is at issue here is an ontological principle that, in Hegelian terminology, asserts the rationality of the real. As Leibniz saw it, the occurrences and characteristics of this world all have a rational explanation, because all the world's arrangements are the product of an all-powerful and totally benevolent rational creator. The

Principle of Sufficient Reason obtains with regard to the contingent arrangements of this world because any breakdown in the explanatory order – any gaps or rents in the fabric of rational explicability – would reflect adversely on the supreme being who has created it. The principle has, for Leibniz, an ultimately theological nature, geared to God's choice of the best of all possible worlds. Leibniz is thus a rationalist of the old school, associated in this regard with medieval scholasticism.

On this basis, we have, from the angle of explanation theory, the principle that there is one fundamental all-explanatory fact that accounts for everything:

$$(\exists g \in F)(\forall f \in F) g \Sigma f$$

Here the variables f, g, etc. range over propositions; F is the domain of facts; and Σ abbreviates the relationship: "provides a satisfactory explanation for." On the medieval and Leibnizian view, such an all-encompassing explanatory principle obtains because one paramount single fact (namely the creation choice of God) ultimately explains *every* fact about the world.

3. Revolution No. 1: Abandoning Totalitarianism

After Leibniz's day, his totalitarian perspective on the rationale of the Principle of Sufficient Reason was soon abandoned in the mainstream tradition of European philosophy. After the Baroque era, the mainstream tradition of European philosophy gave its own twist to the medieval dictum: *non in philosophia recurrere est ad deum*. This major shift was not necessarily motivated by a view of the death of God, but rather reflects a determination not to call upon his services in the context of our explanatory work in science and philosophy. Even theistic theorists inclined to believe that God had more important work to do than help philosophers out of difficulties (as was the case in the systems of Descartes or Berkeley).

Accordingly the totalitarian version of the Principle of Sufficient Reason came to be replaced by the cognate but profoundly different thesis that while every fact indeed has an explanation, this will be something local rather than global. Explanation now became scientifically diversified rather than theologically unitary:

$$(\forall f \in F)(\exists g \in F)g\Sigma f$$

This principle still offers the assurance that everything is explicable. But it no longer calls on one single all-pervasive explanatory consideration to do the job. What we have here is massive shift from *collective* to *distributive* explicability.

4. Revolution No. 2: Abandoning Ultimate Foundations – the Problem of Explanation Stoppers

The shift from a distributive rather than an all-encompassingly collective view of the explicability of things made more urgent than ever the old Aristotelian problem of an explanatory regress. We explain one fact by means of others and these in turn by means of still others, etc. But where does it all end? The older collective view has a simple answer. It all ends in a God who is self-explanatory (*causa sui*, a necessary existent). But with the shift to a distributive view of the explicability of things this convenient synoptic resolution of the problem of explanatory regress became unavailable.

A line of response common from the 17th century onwards is that the end lies in *explanation stoppers*, facts that are discernably of such a nature that they themselves require no further explanation. Theorists of this tendency contemplated a category of *explanation-dispensing explainers* – propositions available for explicatory work that themselves do not *lack* but rather *dispense with* – and indeed leave no room for – further explanation. Over the years, different philosophers have construed the idea of explanation stoppers in somewhat different ways:

- propositions that are self-certifyingly necessary (Spinoza);
- propositions that are explicitly analytic (Leibniz);
- propositions that are necessary a priori (Kant);
- propositions whose denials are inconceivable (Herbert Spencer);
- propositions that are (self-)evident (Roderick Chisholm).

But throughout this tradition, people saw the explanatory regress as coming to a halt in a category of "ultimate" propositions that themselves need no further explanation – a rock-bottom of self-validating propositions that contain their explaining reasons within themselves – within their own substantive nature. We thus encounter such an explanation stopper in f through the circumstance that f is a fact such that:

$$f \Sigma f$$

Any fact f that satisfies this condition is herewith conveniently available for further service as an explained explainer.

However, this explanatory program encounters serious difficulty. To the extent that our beliefs are self-validating they are bound to be either trivial ("Knives have blades") or subjective ("That object looks like an apple to me"). And in neither case are they sufficiently informative to evidentiate any significant substantive conclusion. We are caught in a bind: to whatever extent beliefs are self-validating they are informatively too empty to evidentiate anything objective, and to whatever extent they provide useful evidentiation they are too substantively laden to be self-validating.

Such difficulties of self-evidentiation motivated the shift to a rather different line of approach. They suggested the tactic of seeking for the validating explanation of an explaining fact f not in some particular available fact (conceivably f itself) but rather for an assurance "on general principles" that the fact f is explicable. The guiding idea of this approach is that of looking to facts that are explanatorily available in the context of a particular explanatory problem – facts that "need no explanation" in a way that is not absolute and across-the-board, but rather is variable and context-dependent.

As this approach sees it, in the context of an explanatory question Q that lies before us there will be a (true or factual) general principle P_Q such that

$$P_Q \rightarrow (\exists g \in F) g \Sigma f$$

The situation which now renders a fact f explanatorily available for the resolution of Q (and as itself needing further explanation in the particular Q-coordinated context of discussion) is not that there regress launchingly is some other concrete fact g that explains f, but rather that considerations of general principles assure us at the generic (indefinite) level that an adequate explanation of our explaining fact exists. What is this at issue in providing for usable explainers is thus not a concrete explanation but a generic (indefinite) guarantee. We now have an explanation stopper alright, albeit not one that roots in an across-the-board universalized category of necessary (self-explanatory, self-evident) propositions, but one that roots in a *contextually* delineated category of explanatorily available givens.

The important difference between these two modes of explanation stoppers deserves emphasis. The former sort requires us suppose the availability of (1) particular, concrete, identified propositions which (2) are self-evident (self-justifying). The revised strategy of contextually available explanation stoppers requires *neither* particularly, *nor* self-evidence. It only asks for assurance on the basis of general principles that an explanation is *available* for the given explanation-stopping proposition. In that former case we must know specifically *what* establishes the explanation-stopping proposition, in the second case we need only know *that* the explanation stopping proposition has an explanation of a sort that renders it similarly available. There is a vast difference in specificity. Both modes of justification assure that the proposition at issue is itself explicable (and is thus available for further explanatory work). But in the one case we do not – and *need* not – know what that explanation concretely is; it suffices to be assured on the basis of consideration of general principles that one is forthcoming. It is now possible to achieve explanatory adequacy without requiring a rock-bottom of explanation-stopping self-evidencers. We can avoid an endless explanatory regresses not through to a rock bottom of adamantine necessity, but simply by going back as far as is contextually appropriate within a particular domain of deliberation.

5. REVOLUTION NO. 3: ABANDONING PRETENTIONS TO OMNISCIENCE

(Explanatory Completeness)

But can the demands of reason be satisfied with using materials that are backed by a merely indefinite assurance of explicability rather than by a definite and concrete explanation? Can we conscientiously be content with a merely general assurance of explicability:

$$(\exists g \in F) g \Sigma f?$$

The answer here lies in the recognition that such a procedure is simply inevitable once one acknowledges the absence of an explanatory rock bottom. Once the unsatisfactory nature of a reliance on self-evidence is realized, we have to acknowledge that on any explanatory issue we can always in theory press further, probing ever more deeply into the reason why behind the reason why. And once we realize this, we also have to realize the infeasibility of explanatory completeness. The idea of a *completed* regress has to be abandoned: we cannot insist on going all the way though to the bitter end. We have to come to terms with the fact that the finitude of the human intellect requires us to abandon the identification of *explanatory adequacy* with that of *explanatory completeness*.

We must, after all, recognize and acknowledge the profound difference between what can be managed in theory by an omniscient intellect and what can be managed in practice *by us*, by finite intelligences. In the final analysis, it makes no sense for one to adopt the Leibnizian strategy of the God's eye point of view. And this recognition brings in its wake a profound revision in how one can reasonably construe the demands of reason. At this point we come to what is perhaps the most fundamental of all principles of rationality: *reason is not unreasonable* – it does not stake any demands for what cannot in principle be had. (*Ultra posse nemo – et nihil – obligatur.*)

To clarify this revised approach more thoroghly, let us return to the idea of a contextualistically given guarantee of explicability – a principle P_Q correlative with the context of the explanatory question Q such that:

$$P_Q\Sigma \rightarrow (\exists g \in F) g\Sigma f$$

Given reason's quest for the reason why behind a given reason, it will surely demand an explanatory accounting for the principle P_Q itself? Surely our "explanation" of f remains incomplete unless such an account is given. But then, are we not embarked on an unending regress once more? Let us consider why the answer here is negative.

To validate that explanatory principle, we can resort to a repetition at the higher level of explanatory principles of an argument exactly the same form as that just given above. For in theory there are two ways of proceeding. On the one hand, we can look for a specific and yet deeper principle P^*_Q that validates (accounts for) P_Q – as per:

(1) $P_Q \Sigma P_Q$

And now of course we are en route to a regress showing that at this stop we must worry about the validation of $P\prime_Q$. But a different – already familiar – procedure beckons. For we can also rest satisfied with a generic assurance that, on the basis of general principles,

$$(2) \quad (\exists g \in F)(g \Sigma P_Q)$$

Clearly, in the spirit of our indefinitarian approach this second tactic represents the appropriate stance. The obviously indicated strategy is that of once more having resort to a higher-level explanatory principle that dispenses with particularity. We arrive at a line of reflection that is general-principally all the way through. On this basis, then, a version of the Principle of Sufficient Reason that is altogether abstract rather than concretely reason-specific:

$$(\forall f \in F)(\exists p \in F)(p \rightarrow (\exists g \in F) g \Sigma f)$$

And the crucial aspect of this thesis is that it puts p into the position of depending on f. The status of that pivotal p is now neither specific nor context-free.

But what validates our recourse to this stance? What is it that justifies this new second-level contention itself? Are we not still embroiled in an explanatory regress that can be terminated only at the cost of a resort to the self-explanatory?

6. The Kantian Revolution

Abandoning Substantive Pretentions

The person who shepherded the Principle of Sufficient Reason into the new philosophical order was Immanuel Kant. He too was deeply committed to the principle. As he saw it, not only do we humans insist on finding the reasons for things, but we are committed by the very nature of our intellect to a never-ending search for the reason why behind the reason why. In one key passage he puts the matter as follows:

> The Principle of [Sufficient] Reason is thus properly ... a rule, prescribing a regress in the series of the conditions of given appearances, and forbidding it [viz., reason] to bring the regress to a close by treating anything at which it may arrive as absolutely unconditioned. (CPuR, A508–9 = B536–7.)

As this passage indicates, for Kant the Principle of Sufficient Reason reflects not a learned fact about the world, but rather an inherent postulate of our reason:

> [The] unconditioned is not, indeed, given as being in itself real...; it is however, what alone can complete the series of conditions when we proceed to trace these conditions to their grounds. This is the course which our human reason, by its very nature, leads all of us, even the least reflective, to adopt. (CPuR, A584 = B612.)

For Kant, accordingly, the Principle of Sufficient Reason represents a built-in commitment of our reason to persevere undismayed in the pursuit of an effectively incompletable project: our reason stands committed to the pushing ever onwards in the explanatory project, undeterred by the recognition of its inherent incompletability. Thus Kant maintains:

> [Q]uestions never ceasing – its [reason's] work must always remain incomplete; and it therefore finds itself compelled to resort to principles which overstep all possible empirical employment, and which yet seem so unobjectionable that even ordinary consciousness readily accepts them. But by this procedure human reason precipitates itself into darkness and contradictions; and while it may indeed conjecture that these must be in some way due to concealed errors, it is not in a position to be able to detect them. (CPuR, Aviii.)
>
> Human reason has this peculiar fate that in one species of its knowledge it is burdened by questions which, as prescribed by the very nature of reason itself, it is not able to ignore, but which, as transcending all its powers, it is also not able to answer. (CPuR, Avii.)

For Kant, the principle reflects our reason's commitment to an ongoing process – unceasingly to seek the reason why behind the reason why. It is not a factual guarantee of the *findability* of a sufficient reason, but rather a procedural principle of *seeking* which asserts that our reason, in its inherent commitment to this desideratum, is embarked on a never-ending quest. The situation could not be more different from what we have in Leibniz. For

Leibniz, we know, *a priori*, that every aspect of the world has a cogent explanation. In Kant, however, we have no such cognitive guarantee at all. What we have is a methodological commitment. The ultimate responsibility of the rationality of the real still continues to be attributed to the operation of a mind, but it is now the human mind rather than the mind of God. In sum, the status of the principle is shifted from that of a purported *factual thesis* (which thus requires some sort of evidential substantiation) to a *methodological principle* whose justification can and must be of a different order.

On this basis, the validation of the Principle of Sufficient Reason roots in the Kantian distinction between a definite item of fact that constitutes part of our knowledge and a general principle that regulates how we proceed here. He encapsulates this contrast in the terminology of a constitutive and regulative:

constitutive = we know concretely of p that such and such a situation obtains: $X(p)$

regulative = we take it on the basis of general principles that such a situation obtains: $(\exists p \in F)X(p)$

At the regulative level we have not a concrete fact, but rather a general-principally supposition. In the former case we definitely and concretely know; in the latter we indefinitely and generally presume (suppose, acknowledge). And here there is a crucial epistemological difference for Kant: concrete facts (knowledge claims) must be validated *evidentially* but regulative presumptions can be validated procedurally in terms of their methodological appropriateness.

Accordingly, the preceding formulation of the Principle of Sufficient Reason continues to be the same form as before

$$(\forall f \in F)(\exists p \in F)(p \rightarrow (\exists q \in F)q\Sigma f)$$

But it nevertheless undergoes a great transformation – not a change of *context* but a change of *status*. The principle now takes on a methodological cast, rather than representing a substantive item of knowledge. Its status is not that of a factual thesis but that of a methodological postulate. And as such it does not require the validation of prior reasons but the ex-post-facto retrojustification of its operational utility or essentiality. We avert regress because the principle, being non-factual, does not fall within its own scope of assertion. We are to give up the foundationalistic quest for grounding in something more basic – the "*Begründungsdenken*" criticized by thinkers from Sextus Empiricus to Karl Albert – in favor of a justification by a systemic embedding or in our methodological stance.

The commitment to the rational explicability of the real is, for Kant, an idealized demand of reason. But it is an appropriate (valid) ideal because it plays a useful – nay indispensable – role in the cognitive scheme of things. It challenges us to go even further, to dig even deeper, to improve, not to rest satisfied with the incomplete position we have been able to obtain.

With Kant, then, we find that the Principle of Sufficient Reason undergoes the sea-change from the factual to the methodological order of consideration.

7. Revolution No. 5: Abandoning the Demand for Inferential Explanation

There are two markedly different sorts of explanations – the *inferential* and the *coordinative*.

Consider an illustration. Suppose that it is seen as a "law of nature" that in situations when type S are realized then situations of type S_1 will result in a random 90% of cases and situations of the type S_2 in a random 10% of the cases. Suppose that S is realized in a certain instance and that, in this *particular* instance S_2 ensues. Clearly we cannot explain the occurrence of S_2 through the consideration that, in the circumstance this outcome was foreseeable and only to be expected – because (ex hypothesis) it just wasn't.

The explanation that one *can* give is that inherent in stated facts of the case, viz., that here was an occurrence which, though in the circumstances *improbable*, was nevertheless perfectly normal in its strict conformity to the laws operative in the prevailing circumstances – that the occurrence at issue is perfectly conformable to and consonant with pertinent body of fact and explanatory principle.

In the case of such stochastic (probability-geared) situations we have to see explanation not as involving a necessitarian deduction of facts to be explained but simply as a matter of their smooth accommodation within the relevant fabric of operative law. Our explanatory regress does not trace the thesis to be explained (explanadum) back to some other fact which somehow requires or necessitates it – it is not an *inferential* explanation. Rather it shows how the explained fact fits in smoothly with the explaining facts – it is a *coordinative* explanation. It does not derive the explanatory result from the relevant laws but shows that it fits in smoothly into the fabric of information that they institute.

In this way, the shift from causal (i.e., causally necessitarian) explanation to statistical (probabilistic) explanation carries in its wake a corresponding shift to explain facts and inferentially (by deductive subsumption) but coordinatively (by coordinative integration) into the fabric of relevant law). And this in turn enables a Principle of Sufficient Reason to the effect that every fact has an explanation to be reconstrued in a suitably readjusted way. There need be no regress-involving deduction from yet more fundamental facts when validation can be achieved through the coordination and mutual accommodation of claims.

8. Revolution No. 6: Abandoning Causality

Even as late as Descartes (who spoke of *causa sive ratio*), philosophers did not in general distinguish clearly between *causes* and *reasons*.[3] But Leibniz, to his credit, insisted on maintaining a sharp distinction here (*New Essays*, IV, chap. 17, sec. 1). And such a distinction is, in fact, crucial, because the principles "all facts can be explained" and "every event has a cause" are different sorts of creatures. Given this distinction, it deserves recognition and emphasis that the Principle of Sufficient Reason is actually nowadays in better health than the Principle of Causality.

Consider the thesis:

> Every occurrence (event, development) in nature has a cause: in each and every case there is always a *cause* or explanation that accounts for matters being so rather than otherwise.

It is clear that *this* principle of universal causality is in difficulty. For 20th century physics, with its discovery of stochastic processes that involve the element of "real chance," teaches the breakdown of causality. An unstable atom of a transuranic element (with a half-life, say, of two hours) might decay into other, lighter elements after (say) one second or many hours later. When this decay occurs there is (in the very nature of the situation) no *causal* explanation for its occurring just *then* rather than at some other time. But of course we can perfectly well *explain* the occurrence through the sort of probabilistic explanation that the quantum theory provides. We can give a perfectly good account for it in view of our understanding of nature's *modus operandi*. Explanation remains possible, even though causality breaks down.

It accordingly deserves emphasis that we do not require a Principle of Causality to maintain the Principle of Sufficient Reason. For causal explanation does not exhaust the range of explicability as such. Probabilistic explanations, for example, are still altogether viable. The stochastic occurrence at issue in the preceding example may lack a *cause*, but given the resources of quantum theory we can nevertheless bring it into the domain of rational understanding because it does not lack an *explanation*.

Yet why should we recognize such probabilistic "explanations" as *genuine explanations* – as affording reasons that are not for the phenomena at issue? Why dignify such "weak" nondeterministic inferences with this proud name. Basically for two reasons – one theoretical and one pragmatic. The theoretical reason is that given stochastic rather than deterministic laws, the pattern of reasoning involved in probabilistic explanation is – apart from the differences in the types of laws – strictly parallel to the pattern of reasoning involved in classically causal explanations. We continue to get cogent and informatively helpful answers to questions of the type: Why did *X* rather than *Y* occur? The second, pragmatic consideration lies in the recognition that in contexts where stochastic laws alone are operative, probabilistic explanation is the only type of explanatory reasoning which is possible in the very nature of things. To refuse to accord to such explanatory reasonings the title of "explanation" is to set up so narrow a concept of explanation that many of the reasonings standardly accepted as such in modern scientific discussions are put outside the pale of *explanations proper* by what is – in the final analysis – a fiat of definition buttressed solely by nostalgic memories of what explanation used to be like in good old nineteenth-century physics.

The shift from necessitarian deductions to probabilistic inferences clearly marks another major way-station in the development of the Principle of Sufficient Reason. But there is still more to come.

9. Revolution No. 7: The Turn to Natural Selection

A further important step in the development of thought about the Principle of Sufficient Reason positions it within the framework of the Darwinian revolution. To get at this aspect of the matter, let us consider the situation in the philosophy of Bertrand Russell, the last of the Victorians.

As Russell saw it in one of his later books, the idea that our experience is evidentially indicative of an underlying causal order that provides its explanatory basis is a fundamental presupposition of all of our scientific and common-sensical knowledge of the world:

> That there are such more or less self-determined causal processes [behind our observational experience] is in no degree logically necessary, but is, I think, one of the fundamental postulates of science. It is in virtue of the truth of this postulate – if it is true – that we are able to acquire partial knowledge in spite of our enormous ignorance. (*Human Knowledge: It's Scope and Limits* [New York, 1948], p. 459).

But what validates such a postulate of causal explicability from our observations? We do not *learn* this from experience, because only our commitment to causality makes learning about reality through experience possible in the first place. Rather, its basis – as Russell maintains – is strictly biological:

> Knowledge of connections between facts has its biological origin in animal expectations. An animal which experiences an *A* expects a *B*; when it evolves into a primitive man of science it sums up a number of particular expectations in the statement "*A* causes *B*." It is biologically advantageous to have such expectations as will usually be verified; it is therefore not surprising if the psychological laws governing expectations are, in the main, in conformity with the objective laws governing expected occurrences.
>
> We may state the matter as follows. The physical world has what may be called "habits," i.e., causal laws; the behavior of animals has habits, partly innate, partly acquired. The acquired habits are generated by what I call "animal inference," which occurs where there are the data for an induction... [For] even in animal induction there are elements of validity. The inference from smell to edibility is usually reliable, and no animal makes hay of the absurd inductions which the logician can invent to show that induction is not always valid.
>
> Owing to the world being such as it is, certain occurrences are sometimes, in fact, evidence for certain others; and owing to animals being adapted to their environment, occurrences which are, in fact, evidence of others tend to arouse expectation of those others. By reflecting on this process and refining it, we arrive at the canons of inductive inference. These canons are valid if the world has certain characteristics which we all believe it to have. The inferences made in accordance with these canons are self-confirmatory and are not found to contradict experience. Moreover they lead us to think it probable that we shall have mental habits such as these canons will on the whole justify, since such mental habits will be biologically advantageous. (Ibid. pp. 495–96.)

And so for Russell the Principle of Sufficient Reason, in its explanatory gearing to the causal order of nature, represents a biologically inherent tendency in the human mind that is fixed as a part of the natural programming of its operation through its operation of Darwinian principles of natural selection. Where Kant saw a commitment to the Principle of Sufficient Reason as a procedural commitment inherent in the faculty structure of human reason, Russell sees the Principle of Causality as an evolved instrumentality, developed through biological process on the basis of its evolutionary utility.

To be sure, this view of the matter was not original with Russell, but was already found in C.S. Peirce. Peirce, like Russell, regarded the capacity of the human mind to transport its inductive affairs to root ultimately in the operation of Darwinian natural selection.

> If you carefully consider with an unbiassed mind all the circumstances of the early history of science and all the other facts bearing on the question ... I am quite sure that you must be brought to acknowledge that man's mind has a natural adaptation to imagining correct theories of some kinds, and in particular to correct theories about forces, without some glimmer of which he could not form social ties and consequently could not reproduce his kind. In short, the instincts conducive to assimilation of food, and the instincts conducive to reproduction, must have involved from the beginning certain tendencies to think truly about physics, on the one hand, and about psychics, on the other. It is somehow more than a mere figure of speech to say that nature fecundates the mind of man with ideas which, when those ideas grow up, will resemble their father, Nature. (CP, 5.591 [1903].)

The key here is that the scientist is a member of the species *Homo sapiens* which is the product of an evolutionary development proceeding within nature:

> We now seem launched upon a boundless ocean of possibilities. We have speculations put forth by the greatest masters of physical theorizing of which we can only say that the mere testing of any one of them would occupy a large company of able mathematicians for their whole lives; and that no one such theory seems to have an antecedent probability of being true that exceeds say one chance in a million. When we theorized about molar dynamics we were guided by our instincts. *Those instincts had some tendency to be true;*

because they had been formed under the influence of the very laws that we were investigating. (*Collected Papers*, 7.508 [c. 1898], italics added.)

As this picture of the situation makes clear, these later philosophers agree with Kant in locating responsibility for the regulative commitment to the Principle of Sufficient Reason in the modus operandi of the human mind. They differ, from him, however, in no longer attributing it to the inherently inevitable structure of a given faculty of reason, but rather as a matter of the contingent, biologically developed processes of human cognition.

10. Revolution No. 8: Cultural Pragmatism and the Turn to Rational Selection

The era of Peirce and Russell saw the validation of the Principle of Sufficient Reason take a biological turn. But developments over the past quarter century carry the matter one step further in shifting the burden of accounting for the Principle of Sufficient Reason from a biological to an essentially *sociological* level.

To be sure, the issue can still be viewed in an evolutionary perspective – but with one very important change. For while biological evolution undoubtedly accounts for our *possession* of intelligence, accounting for the way in which we actually go about using it calls for a rather different evolutionary approach – one that addresses the development of thought procedures rather than of the thinkers themselves. Our cognitive capacities and faculties are part of the natural endowment we owe to biological evolution by natural selection. But our cognitive methods, procedures, standards, and techniques are socio-culturally developed resources that evolve through *rational* selection in the process of cultural transmissions through successive generations by teaching, simulation, and the like of those processes and procedures that manifest a purposive efficiency.

This revised approach proposes to see our commitment to the rational explicability of nature's features as the product not of *biological* but of *cultural* evolution – not of natural but of social selection. From this perspective, our endorsement of a Principle of Sufficient Reason is viewed as a methodology-oriented belief – or, perhaps better, an *attitude* – that, having been tried and found helpful, is subsequently transmitted to succeeding generations of intelligent inquirers. Its epistemological status is thus that of a pragmatically validated commitment – one that initially was a working assumption and eventually metamorphosed into a standing presumption through its retrospective revalidation as a successful instrument for furthering the process of inquiry. Its rationality is accordingly such as to give it the status of a pragmatically effective methodological instrumentality to which we become committed by a culturally based process of rational selection rather than a biologically based process of natural selection.

11. Review

The salient idea of the Principle of Sufficient Reason – commitment to the idea that human reason can appropriately demand a rational explanation of the world's facts – played a key role in philosophical thought from the very start of the enterprise. Since the time of the Renaissance it has, however, undergone a series of revolutionary changes. For as we have seen, there has been an ongoing series of diminishing expectation:

- abandoning totalitarianism: rejecting the idea that there need be a single, monolithic principle with inference to which all the world's facts can be explained;
- abandoning ultimate foundations: rejecting the idea that there is a basic bedrock of self-certifying axiomation fact that provides for the explanation of everything;
- abandoning explanatory completeness: rejecting the idea that there is an explanatory rock-bottom, an ultimate terminus of the explanatory regress;

- abandoning claims to a descriptive/substantive status and with it the claim for a descriptively real rather than a methodologically ideal status of the explicability principle;
- abandoning necessitarianism: rejecting the idea that the explaining facts somehow necessitate or predictify the explained facts;
- abandoning causality: rejecting the idea that only causal explanation is adequate explanation;
- abandoning inferentialism: rejecting the idea that explanations must be sequential rather than coordinative;
- abandoning rational inevitability: rejecting the idea that the explicability principle inheres in the nature of the mind (in the very make-up of reason) rather than having the status of a methodological resource that has gained acceptance social acceptance in the community of investigations on the grounds of its utility.

A succession of distinctive stages along a traceable route of development can be discerned here. With Leibniz we have a theocentric position, exactly as in various other great philosophers of his day: Descartes, Spinoza, Berkeley. With Kant we have the Copernican inversion that has the mind of man accomplish work that heretofore was done by the mind of God. With Peirce and Russell we have the Victorian era of philosophy when the philosophical impact of Darwin was at its apogee. With the present generation we reach a stage where philosophers are increasingly inclined to look to the social sciences for their sources of inspiration and correspondingly turn to rational rather than natural selection to validate our commitment to the principle of sufficient reason. Even at a glance, this succession tells a story of domestication, of a descent from Mt. Olympus, of the principle's being reconceptualized from an inescapable imprinting on the human mind to the status of a useful methodological instrument – from hardware to software, from necessity to utility.

What we see throughout is clearly a gradual retreat from absolutism. The course of historical developments maps out an unfolding recognition of human finitude. We here have an historical trend over many centuries that abandons the demand for absolutes and increasingly acknowledges the contingency, complexity, diversity, etc. of the explanatory enterprise.

12. A Puzzling Postscript

The end of our story brings us also to one of the great ironies of intellectual history – the demand of many physicists at the end of the 20th century for one grand unified theory of everything.[4] What these physicists actually want is a theory that restores all of those traditional absolutes: totalitarianism, all-embrancingness, bedrock ultimacy, self-explanatory axiomaticity, necessitarian exigency. Those GUT theorists are willing – nay eager – to trade all those hand-won philosophical victories against absolutism for a nostalgia satisfying pre-modern necessitarianism.

What we have here is a head-to-head battle between a scientific and philosophical tradition that has step by step, worked its way out of a classical absolutism. There is, obviously no safe way of predicting which side will win out – a modernity that abandons absolutism or an idealized mathematical physics that demands it. But there is no room for doubt that a contest of epic proportions is being played out – a Promethean struggle to wrest enlightenment from the reluctant gods. It will be fascinating to see the struggle unfold. The stakes are massive. For what is at issue is not only our scientific view of the world we live in but also our image of the sorts of creatures that we ourselves are.

Notes

1. "Thought will be satisfied with nothing short of showing the *necessity* of its facts, of explaining the existence of its objects, as well as their nature

and qualities." G.W.F. Hegel, *Hegel's Logic* tr. by William Wallace, 3rd ed. (Oxford, 1975), p. 3.

2. "[Scientists and philosophers alike] at each successive step, when they have asked the question Why? have assumed an intelligible answer to be waiting, whether they have found it or not... [Our] rationalism is only a generalization of this assumption." Brand Blanshard, *The Nature of Thought*, 2 vols. (London, 1939); see Vol. 2, p. 476.
3. Compare Spinoza: *Effectus cognitio a cognitione causae dependet et eandem involvit* (*Ethics* I, ax. 4).
4. For a readable account see Steven Weinberg, *Dreams of a Final Theory* (New York, 1993). A more technical exposition is H. M. George, "General Unified Theories" in Paul Davies (ed.), *The New Physics* (Cambridge, 1986).

CHAPTER 2

WHY BE RATIONAL?

Reason requires that we do what is appropriate in matters of belief, evaluation, and action – that we use our intelligence to figure out the proper thing to do and then do it. But why should we do what reason demands? Why heed the demands of reason?

Note this is *not* asking for a validation of reason from without. There can be no such thing. The only sort of validation of anything – reason included – that is worth having is a reasonable one. The validation of rationality is a matter of asking exactly what rationality can say on its own behalf.

What is at issue here is not circularity but simply good sense. Admittedly, the explanation at issue has an *appearance* of vitiating circularity because the force of the validating argumentation with rest on an appeal to rationality: "If you are going to be rational in your beliefs, then you must also act rationally, because it is rational to believe that rational action is optimal in point of goal attainment." But this sort of question begging is simply unavoidable in the circumstances. It is exactly what we want and need. Where else should we look for a *rational* validation of rationality but to reason itself? The overall justification of rationality *must* be reflexive and self-referential. To provide a rationale of rationality is to show that rationality stands in appropriate alignment with the principles of rationality. From the angle of justification, rationality overall is a cyclic process that closes in on itself, not a linear process that ultimately comes to a stop at something outside itself.

There is accordingly no basis for any rational discontent, no room for any dissatisfaction or complaint regarding a "circular" justification of rationality. We would not (should not) want it otherwise. If we bother to want an answer to the question "Why be rational?" at all, it is clearly a *rational* answer that we require. The only sort of justification of anything – rationality included – that is worth having at all is a rational one. The only validation of rationality that can reasonably be asked for must lie in considerations of the systemic self-sufficiency of reason. In the final analysis, "Why be rational?" must be answered with the only rationally appropriate response: "Because rationality itself obliges us to be so." In providing a *rational* justification of rationality – and what other kind would we want? – the best we can do is to follow the essentially circular (but *non-viciously* circular!) line of establishing that reason herself endorses taking this course. Reason's self-recommendation is an important *and necessary* aspect of the legitimation of this enterprise. That presupposition of rationality is not vitiating, not viciously circular, but essential – an unavoidable consequence of the self-sufficiency of cognitive reason. There is simply no satisfactory alternative to using reason in its own defence. This is what is, in the circumstance, what we need and want: circular it may be, but there is nothing improper about it. Already embarked on the sea of rationality, we want such assurance as can now be made available that we have done the right thing.

One may of course quite appropriately ask questions like: Why be rational? Why should I cultivate the truth? Why should I cultivate my best (or true) interests? But in the very act of posing these questions I am asking for reasons – that is, I am evincing my commitment to the project of rationality. Caring for the truth and for one's best interests are simply part and parcel of this commitment. If I do not care for these things, then there is really no point in raising these questions. For in this event I have *already* taken my place outside the precincts of rationality, beyond the reach of reason. At this point there is little more to be said. If I want a reason at all, I must want a *rational* reason. If I care about reasons at all, I am already within the project of rationality. But once I am *within* the project, there is nothing *further* external to reason that can *or need* be said to validate it. At *that* stage rationality is already at hand to provide its own support – it wears its justification on its sleeve. (The project of trying to reason with someone who stands *outside* the range of rationality to convince them to come into its fold is clearly an exercise in pointlessness and futility.)

Irrationality – wishful thinking and self-deception – may be convenient and even, in some degree, psychologically comforting. But it is neither cognitively nor reflectively satisfactory. If it is a viable defense of a position that we want, it is bound to be a rational one. From the justificatory point of view, then, rationality is and must be autonomous. It can be subject to no external authority. Rationality in general is a matter of systematization, and the justification of rationality is correspondingly a matter of systemic self-sufficiency. The only and appropriate way is through the consideration that rationality is in its very nature telelogical and ends-oriented. Cognitive rationality is concerned with achieving true beliefs. Evaluative rationality is concerned with making correct evaluation. Practical rationality is concerned with the effective pursuit of appropriate objectives. Rationality being a telelogical enterprise, its validation does and must proceed in a pragmatic manner with reference to purposive efficacy. It pivots on the question of effectiveness in the context of given ends: of benefits and costs if you will.

What do we stand to lose by going against the demands of reason? Here the matter stands as follows: In ignoring (let alone violating)

<table>
<tr><td>cognitive
evaluative
practical</td><td>rationality we run an
avoidable risk of</td><td>accepting falsehoods
endorsing inferior items
failing to achieve appropriate ends</td></tr>
</table>

In each case we are dealing with an enterprise that has objectives. And in failing to head the standards at issue we are in each case putting the achievement of these objectives at risk. Accordingly the *validation of rationality consists in the consideration that its violation would compromise the successful pursuit of appropriate ends*. And this argumentation is itself a paradigm example of a pragmatic/teleological validation.

Note that all of rationality is engaged in this process of validating reason. It pivots on there being (cognitively) sound reason to think that an (evaluatively) appropriate end of the enterprise will become less likely to be (practically) realized. We have here a line of deliberation in which all of the sectors of rationality are at work. But in its overall nature the justification is *pragmatic* by way of pivoting on the issue of purpose realization.

And there is good reason for this pragmatic orientation. For a purely theoretical defense of rationality that demonstrates that logically incoherent is simply unavailable.

How would one logically refute those who set rationality at naught? How can we argue against the sceptical nihilist who refuses to accept any and all truth claims or the psychopath who has a totally weird system of values and priorities or the madman whose actions deliberately frustrate the realization of his own ends?

Take the sceptic who rejects all of the usual ground rules of epistemic practice and refuses to accept anything whatsoever. How can one possibly *reason* against him – argue him out of his position? All rational argumentation proceeds from premisses, from something accepted (at least *pro tem*). One who rejects everything – accepts nothing – obviously allows no fulcrum for the lever of reasoning.

Or, take the nihilist who rejects all the usual values that people hold and – in particular – is psychopath also persists in setting the value of other people and their interests at nought. Such a person is beyond the reach of moral deliberation because matters in this domain can only be addressed *rationally* on the basis of standards – that is, by proceeding from an evaluative standpoint vis-à-vis personhood. The value nihilist clearly places himself outside the reach of rational appeal and moral suasion.

Again, take the madman whose actions frustrate his own aims. His failure is not – or need not be – one of logical reasoning. At each step of a dialectic of argumentation he would well come up with some weird story to rationalize his actions. It's just that the whole Rube Goldberg contraption doesn't make any sense.

We have no alternative but to acknowledge that we cannot refute such rationality rejectors by *purely* logico-theoretical reasoning. Those who turn their back on the principles of reason are for this very reason secure from defeat by reason's means. For we will then be unable as a matter of course to get a foothold for logico-theoretical argumentation to get under way.

In opposing nihilism and irrationalism in its various forms we have to proceed in the mode of practical rather than logico-theoretical reason. The "inconsistency" in which these people are caught up lies in this practical rather than theoretical domain. For the crux is that in abandoning rationality we discernably compromise the chances of accomplishing our own ends – those to which we commit ourselves upon entering into the theoretical and practical projects at issue.

We cannot reasonably set it up as a task for the rational enterprise to discharge the in principle impossible mission of convincing an (otherwise rational) moral nihilist (psychopath) to walk in the paths of morality and benevolence or to convince an (otherwise rational) epistemic nihilist and all-out sceptic to walk in the paths of ordinary rational inquiry. In all such cases, the aim of the enterprise must be conceived of as involving no more (though also no less) that of convincing otherwise rational and normal (commonsensical) people that they should not become psychopaths or sceptics in the first place, since doing so would involve costs that yield no compensating benefits – that for the angle of such (normal, common-sensical) people as we generally find in the world about us there is no suitable balance of advantages gained over benefit foregone in being sceptical or psychopathic.

And this task is not all that difficult. The sceptic says "I have such high epistemic standards that nothing can meet them." But what does this gain him? Nothing: he leaves the cognitive playing field altogether empty handed with all questions unanswered. The psychopath says "I set the value of others and their interests at nought." Now this does indeed gain him *something*, namely the advantage of not having to bother about others. But it gives this advantage at the expense of the resulting isolation and loss of bonds of affection and congeniality that exacts a horrendous price in terms of alienation and misery – any rate for normal people.

In the only way in which it makes sense to be concerned about the question "Why be rational?," then, there is a perfectly good and satisfactory response. It makes sense in the manner of all practical reasoning, to wit, that failing to do this produces significant losses not compensated for by preponderating benefits. Rationality has the perfectly rational justification that in failing to heed the dictates of reason we came up on the short end of the balance of benefits gained versus advantages foregone.

The aim of the enterprise of grappling with the question "Why be rational?" is not, however, to "refute the sceptic" or "refute the psychopath" by some categorical demonstration that this position is logically untenable, unavoidable to any rational person who is prepared to maintain the position come what may and simply accept whatever negativities this involves. Rather the aim is to point out that taking this position calls for paying a price that *sensible* people (*l'homme moyen rational*) would be unwilling to pay because the cost/benefit, gain/loss balance stands too adversely. We accept the guidance of reason because rational considerations indicate that this affords us the *best overall chances* of reaching our goals. We act, in short, on the basis of *faute de mieux* considerations, of "this or nothing better – as far as the eye reaches."

The fact of the matter is that we cannot *prove* that rationality pays. We do not know that acting rationally in the particular case at hand will *actually* pay off – nor can we even claim with unalloyed assurance that it will *probably* do so (with real likelihood rather than subjective probability). We can only say that, as best we can judge the matter, it represents the most promising course at our disposal. We have no guarantees – no means are at our disposal for pre-establishing that following rationality's counsel actually pays. All that we know is that being rational in the case at hand is the (rationally) best that we can do in the circumstance.

We recognize full well that even the most rationally laid scheme can misfire. Reality is not always and inevitably on the side of the strongest arguments. And in any case we see into the future but dimly. Reason affords no categorical guarantee of success, but only the reassurance of having taken the best rational step – of having done as well as one could in the circumstances of the case. One cannot say, flatly and unqualifiedly: "You should be rational because rationality pays in rendering success if not certain then at any rate more probable." Rather, we have to content ourselves with: "You should be rational because this affords the best rationally foreseeable prospects of success – on the whole and in the long run."

And so, we follow reason because this makes good rational sense to do so, seeing that it affords us with the best visible prospect for realizing our objectives. One should be rational in general for just the same sort of reason as one should be rational in the specific case of the hungry man's choice between eating bread or sand – namely, that by all available indications this course represents the most promising prospect of attaining one's sensible goals.[1]

Throughout our cognitive and practical affairs we have to conduct our operations under conditions of risk. And so, when we do the rational thing but it just does not pan out, we simply have to "grin and bear it." Rationality affords no guarantees. By the very nature of what is involved in rational procedure, the voice of reason speaks. The matter is one of sensibly calculated risks and plausibly expectable benefits. But this too may still be cold comfort when things go wrong. Then, all we have is the satisfaction of having done our best. The long and short of the matter is the nothing "obliges" us to be rational except our rationality itself.

Of course, one may somehow *prefer* not to be rational. With belief, I may prefer congeniality to truth. With action, I may prefer convenience to optimality. With value, I may prefer the pleasingly base to the more austere better. On all sides, I may willfully opt for "what I simply like," rather than for that which is normatively appropriate. But if I do this, I lose sight of the actual ends of the cognitive, practical, and evaluative enterprises, to the detriment of my *real* (as opposed to *apparent*) interests. It lies in the nature of things that reason is on the side of rationality. To be sure, she offers us no guarantees. Yet, if we abandon reason there is no place better that we can (rationally) go.[2]

NOTES

1. Compare René Descartes, *Discourse on Method*, sect. iii, maxim 2.
2. Some of the themes of this discussion are further developed in Chap. 3, "The Rationale of Rationality," of the author's *Rationality* (Oxford, 1988).

CHAPTER 3

REASON AND REALITY

1. INTRODUCTION

It is an unavoidable fact that substantial difficulties confront human inquiry in its project of gaining an appropriate understanding of nature. In this connection, the present discussion will examine the ramifications and implications of four fundamental considerations regarding the situation of scientific theorizing:

- data underdetermine theories;
- theories underdetermine facts;
- reality transcends the *descriptive* resources of language (and symbolic representation in general);
- reality transcends the *explanatory* resources of science.

Let us consider, step by step, each one of these four sceptical considerations which, taken together, spell the inevitable imperfection of our attempts to domesticate reality to the conditions of human understanding.

2. DATA UNDERDETERMINE THEORIES

One of the most basic lessons of modern epistemology is that the available observational data underdetermine theories. For observations are always discrete, finitely enumerable, and episodic. They deal in specific information-yielding episodes that transpire at particular spatiotemporal locations. Scientific theories on the other hand are assertorically general and non-finite: they deal in how certain features of a generic sort characterize situations of a certain kind always and everywhere. Accordingly, theories transcend the data and reach beyond them, so that there is always an evidential gap between the claims of a theory and the particular facts that afforded its supporting data. (The fitting of a continuous curve to discrete data points provides a somewhat oversimplified analogy.) Theories – as Pierre Duhem long ago emphasized in his classic book on *The Aim and Structure of Physical Theories* (1905)[1] – lie beyond the possibility of definitive confirmation by data.

This data-transcendence of all theoretical claims means that there will always be various alternative ways of rounding the data off into generalized theories. The data themselves can never constrain unique theoretical resolutions. Observed data are finite; possible hypotheses infinite; and the finite cannot constrain the infinite. The concrete realities we experience observationally are always to some extent ambiguous. They always admit, in principle, of alternative theoretical rationalizations. Theorizing without data is futile. But even theorizing on the basis of data involves projection and interpretation, and thus carries in its wake the prospect of ambiguity, diversity, and discord. The experiential realities determine and canalize the work of theorizing reason but they cannot impose altogether unique resolutions upon it. In sum, theories are always informatively underdetermined by data. Moreover –

3. Theories Underdetermine Facts

Any precisely formulated (logico-mathematically formalized) theory admits of a variety of concrete actualizations: formalized theories can always be represented by a variety of distinct realizations or models. For any scientific theory that is formalized in a system of propositions articulated through the instrumentalities of logic and mathematics confronts a stark fact. As the celebrated theorem of Löwenheim-Skolem shows, any set of propositions framed in the formal language of mathematics (i.e., logic and set theory) can be fitted out with a purely mathematical model – more specifically, can in fact be reconceptualized as presenting merely a set of facts about natural numbers. Thus no such theory can ever have enough substantive content to constrain any one particular intended physical interpretation. The fit of a formalized empirical theory to the reality it purports to characterize can never be made so tight as to preclude alternative, equally valid representations in totally different areas of discourse. No theory can uniquely dictate its own interpretation. In sum, formalized theories can never be fact-specific – by their very nature they have an aura of ambiguity, leaving open a range of alternatives regarding the facts they manage to capture. Accordingly, just as facts underdetermine theories, so do theories underdetermine facts.[2]

4. Reality Transcends the Descriptive Resources of Language

Our knowledge regarding the world's things is always cast in a linguistic/systematic frame of reference. Now in this context it is important to distinguish between a *truth* and a *fact*. A "truth" is something that has to be understood in *linguistic* terms – the representation of a fact through its statement in some actual language. Any correct statement in some actual language formulates a truth. (And the converse obtains as well: a *truth* must be encapsulated in a statement, and cannot exist without linguistic embodiment.) A "fact," on the other hand, is not a linguistic entity at all, but an actual circumstance that exists objectively in its own right – an aspect of the world's state of affairs. And once again, the same principle is operative: statements are denumerable, facts potentially nondenumerable; and (to reemphasize) the lesser cannot constrain the larger. Theories are language bound, but facts outrun these linguistic limits.

The real in all its blooming buzzing activity is too rich for adequate representation by the recursive and enumerable resources of our language. And so the entities as theory describes them are never more than rough approximations to the entities that comprise the real world.

Moreover, any adequate account of inquiry must accordingly recognize that the ongoing process of information acquisition at issue in science is a process of *conceptual* innovation, which always leaves certain facts about things wholly outside the cognitive range of the inquirers of any particular period. Caesar did not know – and in the then extant state of the cognitive art could not have known – that his sword contained tungsten and carbon. There will always be facts about a thing that we do not *know* because we cannot even *conceive* of them in the prevailing conceptual order of things. To grasp such a fact means taking a perspective of consideration that as yet we simply do not have, since the state of knowledge (or purported knowledge) has not reached a point at which such a consideration is feasible. Any adequate world-view must recognize that the ongoing progress of scientific inquiry is a process of *conceptual* innovation that always leaves various facts about the things of this world wholly outside the cognitive range of the inquirers of any particular period.

The properties of a thing are literally open-ended: we can always discover more of them. Even if we were (surely mistakenly) to view the world as inherently finitistic – espousing a Keynesian principle of "limited variety" to the effect that nature can be portrayed descriptively with the materials of a finite taxonomic scheme – there will still be no *a priori* guarantee that the progress of science will not lead *ad indefinitum* to changes of mind regarding this finite register of descriptive materials. And this conforms exactly to our expectation in these matters. For where the real things of the world are concerned, we not only expect to learn more about them in the course of scientific inquiry – to discover new and different features and facets. Accordingly: *we expect to have to change our minds about their nature and modes of comportment.*

The language of emergence can perhaps be deployed usefully to make the point. But what is at issue is not an emergence of *the features of things*, but an emergence in our *knowledge* about them. Blood circulated in the human body well before Harvey; substances containing uranium were radioactive before Becquerel. The emergence at issue relates to our cognitive mechanisms of conceptualization, not to the *objects* of our consideration in and of themselves. Real-world objects must be conceived of as being what they are antecedently to any cognitive interaction – as being there right along, "pregiven" as Edmund Husserl put it. All cognitive changes or innovations – on the other hand – are to be conceptualized as something that occurs on *our* side of the cognitive transaction, and not on the side of the *objects* with which we deal.[3]

As these considerations indicate, reality outruns the descriptive resources of language. Even if Wittgenstein was right and the limits of knowledge are the limits of *our* world, they are nevertheless not the limits of *the* world. We do and must recognize the limitations of our cognition. We cannot justifiably equate reality with what can, in principle, be known by us, nor equate reality with what can, in principle, be expressed by our language. And what is true here for the situation of our sort of mind is true for any other sort of finite mind as well. Any physically realizable sort of cognizing being can know only a part or aspect of the real.

5. Reality Outruns the Explanatory Resources of Scientific Theorizing

Not only does reality outrun the *descriptive* resources of language, but it outruns the *explanatory* resources of scientific theorizing as well. The basic circumstance at issue in this regard was already apprehended by Aristotle at the very outset of the scientific enterprise. It pivots on two considerations:

- Scientific explanation proceeds at the level of generality. It deploys generalizations which have the form of stating how *all instances* of a particular natural kind comport themselves. It never descends below the level of minimal natural kinds (*infima species*) to discuss the comportment of individuals;
- Concrete things, situations, and processes always have particular individualized features (Aristotle called them "accidents") that we can confront in experiences but cannot capture in a science that addresses generalities alone.

It clearly follows from these two considerations that the concrete particularities that we encounter in experience always have features of concreteness and detail that escape the abstract and general concept mechanisms that science puts at our disposal. Science lays down the general rules, but the concrete chess game that nature plays under their aegis involves detail that is not fully provided for in the generalities at issue. The particularistic/experiential aspect of the real – its concreteness as exhibited in our spatiotemporal interaction with it – will always exhibit features that eludes the explanatory grasp of generalizing science. This state of affairs is brought home to us in a particularly vivid way in modern quantum theory. For quantum theory maps out a probabilistically/statistically structured spectrum of possibilities whose concrete details are nowise settled but left open to the working-out of the historically developmental processes of nature. The actual concrete experiences that the course of nature provides for (the quantum *observations* that we make) involve a level of particularity to which quantum *theory* is unwilling and unable to descend. (A "concretization of probabilistically structured possibilities" is no more than a not infelicitous characterization of "the collapse of the wave packet.") Insofar as an analysis along these lines is right, then, the quantum theory illustrates in a particularly vivid way that the concrete particularities of natural *history* reach beyond the abstract and generalized rules of natural *science*. The ontological detail of the concretely real is always greater than the range of what can be accounted for by the theorizing resources of science. The philosophical lesson here is that nature's actual states of affairs have features that cannot be descriptively captured by – nor explanatorily accounted for in – the resources of quantum theory.

6. THE BURDENS OF COMPLEXITY

The overall lesson of these deliberations is that the extent to which the real is rational – the extent to which the human mind can get an adequate grip on reality – is unavoidably limited. The fact is that the complexity of the real outruns the limits of our powers – in this or any other attainable states of the cognitive art. We have to face up to the idea of a world that is too complex for the descriptive resources of our language and the explanatory resources of our science.

The physicist David Bohm has maintained along these lines that, "at least as a working hypothesis science assumes the infinity of nature; and this assumption fits the facts much better than any other point of view that we know."[4] And in line with this approach, he postulates a principle of unending complexity in nature's make-up. Such an infinity-of-nature postulate may take various forms. One of them involves the suggestion that the domain of natural laws is inexhaustible, following the lines of Pascal's idea that nature might be an endless nest of Chinese boxes, with microscopic worlds emplaced within ever more minute microcosms.[5] It contemplates a system of wheels within wheels – somewhat on analogy with the idea that every organism comes equipped with a yet smaller organism to bite it. Sometimes this shrinkage toward the submicroscopically small is inverted into an expanding sequence moving unendingly upward toward the supratelescopically large – emplacing macroscopic worlds within ever more macroscopic ones. (Perhaps what we see as a galaxy is simply a subatomic particle of a macroworld, and our whole universe no more than an atom.) Following this line of thought, one might project an ascending sequence: subatomic particles, atoms, molecules, organic microorganisms, animals, populations, life-systems, solar systems, galaxies, galactic clusters, worlds, world-systems, and so on, each furnished with its own characteristic modes of operation. Every system is potentially a member of a larger system. On each view – with ongoing increase or decrease alike – the prospect of a theoretically unending progress of science is assured by the existence of an unending series of new worlds embracing (or embraced by) the old.[6] Science is potentially unending because there are ever new orders of nature to be explored.

Some theorists – C. S. Peirce, for example – see the opening up of new realms of phenomena in *evolutionary* terms. Early on in world history, before the evolution of complex molecules, there was no place for biological laws; in the era of Neanderthal man there was no room for political economy. As the cosmos grows older, new modes of natural organization gradually evolve to afford new phenomena that are governed by emerging laws of their own – laws which previously had no opportunity to come into operation. There is progression from laws of individual physical particles to laws of increasingly elaborate organized complexes thereof. Since the universe affords a varied panorama of modalities of physical process evolving over time, a science that reflects this will continue to find new grist for its mill.

Still other theorists consider the element of chance and randomness at work in the world so as to provide a potentially unending hierarchical order of processes governed by principles that represent statistical summaries, as it were, of successively emergent tendencies. In this vein, the French physicist Jean Paul Vigier contemplates a succession of lawful orders conceived along Hegelian dialectical lines, reasoning as follows:

> At all levels of Nature you have a mixture of causal and statistical laws (which comes from deeper or external processes). As you progress from one level to

another you get new qualitative laws. Causal laws at one level can result from averages of statistical behavior at a deeper level, which in turn can be explained by deeper causal behavior, and so on *ad infinitum*. If you then admit that Nature is infinitely complex and that in consequence no final stage of knowledge can be reached, you see that at any stage of scientific knowledge causal and probability laws are necessary to describe the behavior of any phenomenon, and that any phenomenon is a combination of causal and random properties inextricably woven with one another.[7]

Such an approach envisions an unending hierarchy not of physical structures but of modes of order characterizable through statistical patterns in natural phenomena – and of the causal laws that result from applying the laws of large numbers to extract stabilities from such statistics.

There are two in principle quite different sorts of complexity hierarchies. One kind is a hierarchy of material systems related by physical inclusion: particles, atoms, molecules, macrolevel physical objects, stars and planets, galaxies, galactic clashes, etc. Or again: molecules, cells, organs, organisms, colonies, etc. However, such *physical* hierarchies of compositional structure stand in contrast to *law* hierarchies of physical order with stages moving from base-level laws that govern phenomena to higher-level laws coordinating laws, and upwards to laws coordinating laws that coordinate laws, and so on. The latter sort of hierarchy of lawful order does not presuppose the former. It can go on indefinitely even in a world whose structural complexity depth is finite: complexity of operation does not require complexity in point of physical constitution. Even a world that is finite in the structural complexity of its physical constitution may well exhibit a "hierarchy of nomic orders" with an ongoing sequence of levels of higher-order laws.

In a law hierarchy, any law is, potentially, a member of a wider family of laws which will itself exhibit some lawful characteristics and thus be subject to synthesis under still "higher" laws. We thus move from laws governing phenomena (first order laws) to laws governing such laws (second order laws) and so on, ascending to new levels of sophistication and comparative complexity as we move along. No matter what law may be at issue, we can ask questions about it that demand an answer in lawful terms. Given such an unending exfoliation of law levels, new metadisciplines can in theory always spring up to relate old disciplines.

A fundamental principle is at work here. Since laws take the hypothetical form "If such-and-such possibilities were realized, such-and-such results would ensue," the finiteness of the manifold of actuality does not entail the finitude of the diversity of laws. Laws, being hypothetical in nature, belong to the domain of possibility, and this domain is not necessarily finitized by the finitude of the actual.[8]

It is clear that such an infinite proliferation of laws would also serve to block any prospect of completing science. There is thus no need to suppose that the *physical* complexity of nature is unlimited for nature to have an unlimited *cognitive* depth. After all, the prime task of science lies in discovering the laws of nature, and the ongoing law complexity of nature suffices for our present purposes of providing for potentially endless discovery.

A supposition of the *structural infinitude* of nature is accordingly not even needed to assure nonterminating progress in science. An unending depth in the *nomic complexity* of the world's legal code would be quite enough to underwrite the potential limitlessness of scientific innovation.

But one can go still further.

Even a system that is finitely complex *both* in its physical make-up *and* in its basic law structure might yet be infinitely complex in its actual *operations* over time. For the operations of a structurally finite and nomically finite system can yet exhibit an infinite intricacy in *operational or functional complexity*, manifesting this limitless diversity in the working out of its processes rather than at the spatio-structural or nomic level. Even were the number of constituents of nature to be small, the ways in which they can be combined to yield products in space-time might yet be infinite. Think here of the examples of letters/syllables/words/sentences/paragraphs/books/genres (novels, reference books, etc.)/libraries/library systems. Even a finite nature can, like a typewriter with a limited keyboard, yield a body of text that admits of potentially endless variation. It can produce a steady stream of new phenomena – "new" not necessarily in kind but in their functional interrelationships and thus in their theoretical implications – on this basis of which our knowledge of nature's operations is continually enhanced and deepened.

There is no need to assume a "ceiling" to such a sequence of levels of integrative complexity of phenomenal order. The different levels each exhibit an order of their own. The phenomena we attain at the *n*th level can have features whose investigation takes us to the $(n + 1)$th. New phenomena and new laws presumably arise at every level of integrative order. The different facts of nature can generate conceptually new strata of operation that yield a potentially unending sequence of levels, each giving rise to its own characteristic principles of organization, themselves quite unpredictable from the standpoint of the other levels. Even a world that is relatively simple as regards its basic operations can come to have an effectively infinite *cognitive* depth, once one proceeds to broaden one's notion of a natural phenomenon to include not just the processes themselves and the products they produce but also the *relationships* among them.

Consider, for example, some repeatedly exemplified physical concept and contemplate the sequence of 0's and 1's projected according to the rule that the $x_i = 1$ if the concept applies (or is exemplified) on occasion number i, and $x_i = 0$ if the concept does not apply on this occasion. Whenever two such feature concepts, C and C*ı*, generate such sequences, say

C: 0100110100 ... C
C*ı*: 1001011010 ...

we can introduce the corresponding *matching sequence* for C and C/, 0010010001 ..., which is such that its *i*-th position is 1 if the two base sequences *agree* at their respective *i*-th positions, and is 0 if they *disagree*. Such matching sequences will have a life of their own. For example, even if two base sequences are random, their matching sequences need not be – for example when those base sequences simply exchange 0's and 1's. (Even random phenomena can be related by laws of coordination.)

Note, moreover, that one can always regard matching sequences themselves as further base sequences, so as to yield "second-order" phenomena, as it were. One can then proceed to examine the relationships between them – or between them and other base sequences. This process yields a potential hierarchy of "laws of coordination" – at level $i + 1$ we have the laws of coordination between sequences at level *i*. Such a perspective illustrates how simple base phenomena can ramify to bring more and more grist to the mill of study and analysis. Increasingly sophisticated mechanisms of conceptual coordination can lead us to regard the same phenomena in the light of different complexity levels. Quite different regularities and laws can emerge at different levels.

And there is no reason why this sort of thing cannot go on and on. For the system always exhibits new patterns of phenomenal order over time, and so there is always more to be learned about it. There will always be new levels of functional complexity of operation to be investigated with such a system. It will always be possible to discern yet further levels of structured relationship.

Even when we continue to concern ourselves with the same species of object (e.g., symbols), and the same basic laws (e.g., combination rules or grammars), we can nevertheless introduce increasingly powerful means of analysis to secure new sorts of data. Consider an analogy. The chess master and the beginner make exactly the same sorts of moves – the individual pieces behave exactly alike for each of them. At this level their realms "are governed by exactly the same laws." An observer exclaims:

"I know all about chess. For I have now discovered the rules according to which those pieces move." Splendid! But also naive, since this business of "the basic rules of the game" is of course merely Step One. It is in point of the complexity of their governing principles that masters and beginners differ decisively.

To be sure, that chessboard is limited and finite – and so are the pieces sitting upon it. Moreover, so are the basic "laws of nature" – the moves that the different pieces may make. Nevertheless, the moves of a game of chess are covered by very different categories of "rules" – not just "the rules of the game" that define those basic legal chess moves, but rules of tactics and deeper level principles of strategy. The game can be played – and thus studied – at very different levels of depth or sophistication.

As the analogy of chess makes clear, it is radically mistaken to think that we have gotten to the bottom of things when we merely grasp nature's basic physical laws. The question of what sort of game she is playing by those rules – those *basic* laws – remains open. Even though nature might be of finite physical and nomic complexity as regards it physical structure and its basic procedural laws, nevertheless it could be infinitely diverse in the unfolding operational complexity of its phenomenal products over time. To further our understanding of the world about us we need departments of biology and economics as well as departments of physics.

Regardless of the known character of a sequence of phenomena that we confront, we can never rule out the possibility that yet further patterns of relationship exist. For there will be patterns of phenomena, and patterns of such patterns, and patterns of patterns of such patterns – and on and on. We can study letter sequences as such, or move on to the level of words, and thence to sentences, and thence to paragraphs, and so to chapters, to books, to book categories, to book systems (French vs. Chinese literature), etc. Every new level of consideration will afford phenomena of its own that will themselves admit of further study and analysis.

Confronted with repetitive phenomena of any description, inquiry can always in principle find new grist for its mill.

In theory it may occur that just the same relationship patterns simply recur from level to level – that the patterns of phenomena that we encounter at level $i + 1$ simply reduplicate those already met with at level i. (When it is physical configurations we are dealing with, then this situation of pattern repetition at different levels of scale will yield the "fractal" structures made prominent by E. Mandelbrot.) However, this is a very special case that certainly does not obtain across the board. We have no good reason to think that our world is fractal in the structure of its processes.

But is natural science not bound by a Principle of Simplicity – is it not committed to the idea that nature proceeds in fundamentally simple ways? By no means! We have no ground whatever for supposing the "simplicity" of nature. The so-called "Principle of Simplicity" is really a principle of complexity management:

> Feel free to introduce complexity in your efforts to describe and explain nature's ways. But only when and where it is really needed. Insofar as possible "keep it simple!" Only introduce as much complexity as you really need for your scientific purposes of description, explanation, prediction, and control.

Such an approach is eminently sensible. But of course such a principle is no more than a methodological rule of procedure for managing our cognitive affairs. Nothing entitles us to transmute this methodological precept into a descriptive/ontological claim to the effect that nature is finitely complex.

But there is yet another, more fundamental, aspect to the matter. The usual recourse to an infinity-of-nature principle reflects a strictly one-sided approach that places the burden of responsibility for the incompletability of science solely on the shoulders of nature herself. The potential endlessness of scientific progress is taken requires limitlessness on the side of its *objects*, so that the infinitude of nature must be postulated at either the structural or the nomic and/or operational level. But this approach is a mistake. For what really matters for ongoing scientific innovation is *cognitive* rather than *physical* complexity. Scientific innovation – the cognitive exploration of the ways of the world – pivots on the *interaction* of the mind with nature, on the *mind's exploitation of the data to which it gains access* for the sake of penetrating the "secrets of nature." The crucial fact is that scientific progress hinges not just on the make-up of nature herself, but also on the character of the information-acquiring processes by which we investigators investigate it. Responsibility for the *cognitive* inexhaustibility of nature need not lie on the side of nature at all, but can in principle rest one-sidedly with us, its explorers.

As the idea of an inwards spiral shows, even though the distance that separates us from a destination is finite, there nevertheless remains the prospect of having to confront the prospect of an infinite journey to it – of constant progress towards the goal without even attaining it. (And who is to say that those difficult later steps are inherent no more important than the easier early ones?)

Suppose a natural system to be such that a certain parameter p cannot be evaluated by us at the precise time-point t – for essentially technical reasons – but only on average during an interval around t. In such a case, the system can be very simple indeed – it need contain *no* complexities apart from those required to assure the preceding assumptions – and yet the prospect of endless cognitive progress is nevertheless available. For as our capacity to make p-determinations down to smaller and smaller time intervals increases from minutes to seconds to milliseconds to microseconds, and so on, we can obtain an increasingly comprehensive insight into the *modus operandi* of the system and can obtain ever fuller information about it that could not have been pre-determined on the basis of earlier knowledge. Since averages at levels of larger scale do not determine those at smaller ones, quite different modes of comportment – and thus laws – could manifest themselves at different levels.

Again, suppose a (totally random) sequence of 2's and 3's, on the order of 2 3 2 2 3... and suppose further a transformation that substitutes the pair 10 for 2 and 11 for 3 so as to yield: 10 11 10 10 11... We are now in a position to ascertain such "laws" as:

(1) The sequence of *even-numbered* positions 01001... will be a random mix of 0's and 1's (which simply mirrors the initial random sequence of 2's and 3's);
(2) All the *odd-numbered* positions are uniformly filled by 1's.

We confront a peculiar mixture of randomness with regularity here. But of course it is only by studying *pairs* in that initial zero-one sequence that we can discern its code – the study of statistics that ignores patterns of periodicity will not bring regularities of the sort at issue to our notice. Only by bringing appropriate coordinative *concepts* to bear can we discern the "laws" at issue. And no matter how far we push forward along such lines, the prospect of *further* positional laws can never be eliminated – nor downplayed by claiming that such laws are inherently less significant than the rest.

It becomes crucial in this context that higher-level patterns are not necessarily predictable from lower-level ones. Knowing the frequency with which individual letters like A and T occur in a text will not tell us much about the frequency with which a combination like AT occurs. When we change the purview of our conceptual horizons, there is always in principle more to be learned – novelty that could not have been predicted from earlier, lower-level information.

The introduction of *new conceptual perspectives* is the key. Each time a new discipline opens up its novel characteristic perspective, a vantage point is gained for the reappraisal of old issues in other fields. (The body of Shakespeare's work is finite, but this does not mean that Shakespeare scholarship will come to the end of the road. Think of how psychoanalysis, for example, or sociology – disciplines that did not exist in the Bard's own day – have been used as launching platforms for innovations in Shakespearean studies.)

The contemplation of such examples shows that the complexity of exfoliating hierarchy levels need not inhere in the structural make-up of nature herself, but may derive from the conceptual mechanisms being deployed to study it at ever greater depths of sophistication. The case is similar to that of the geometer, whose hierarchy of definitions, axiomatic facts, lemmas, theorems, and so on, does not reside in the geometer's materials as such, but in the conceptual taxonomy that he chooses, for cognitive reasons, to impose on these materials. By herself, nature presumably lacks "depth" altogether. For depth (like difficulty) is an inherently relative matter, generated through the operation of a cognitive perspective. The endless levels at issue will not be *physical* levels but *levels of consideration* that emerge from the *modus operandi* of inquiring beings. Complexity, after all, lies less in the objects than in the eyes of their beholder. (As Herschel ruminated long ago, particles moving in mutual gravitational interaction are, as we human investigators see it, forever solving differential equations which, if written out in full, might circle the earth.[9]

If we are sufficiently myopic, then, even when the scene we examine is itself only finitely complex, an ever ampler view of it will emerge as the resolving power of our conceptual and observational instruments is increased. When we make measurements at a given level of accuracy, the world may appear X-wise; and when that accuracy is improved by an order of magnitude it may appear Y-wise. At each successive state-of-the-art stage of increased precision in our investigative proceedings, the world may take on a very different nomic appearance, not because it *changes*, but simply because at each stage it *presents* itself differently to us.

Accordingly, the reason for the cognitive inexhaustibility of nature need not rest with nature alone. The character of our information-gathering procedures, as channeled through our conceptual/theoretical perspectives, is also bound to play no less crucial a part.[10] Innovation on the side of data can generate new theories, and new theories can transform the very meaning of the old data. This dialectical process of successive feedback has no inherent limits, and suffices of itself to underwrite a prospect of ongoing innovation in the study of nature.

Any adequate theory of inquiry must recognize that the ongoing process of science is a process of *conceptual* innovation that always leaves certain theses wholly outside the cognitive range of the inquirers of any particular period. This means that there will always be facts (or plausible candidate facts) about a thing that we do not *know* because we cannot even *conceive* of them. For to grasp such a fact means taking a perspective of consideration that we simply do not have, since the state of knowledge (or purported knowledge) is not yet advanced to a point at which its entertainment is feasible. In bringing conceptual innovation about, cognitive progress makes it possible to consider new possibilities that were heretofore conceptually inaccessible.

An inferior mechanism for the accommodation of cognitive complexity cannot satisfactorily accommodate a higher degree thereof. We cannot adequately translate Shakespeare into Pigin English. We cannot satisfactorily make do with five questions in a guessing game where even Twenty Questions are only just sufficient. An understanding that operates at a lesser level of complexity cannot adequately grasp an environment that operates at a greater. There is much to be said for an epistemic humility which acknowledges that in any state of the cognitive art our knowledge must be seen as an imperfect and inadequate situation to the problem of depicting "the way it really is" as regards nature's *modus operandi*.

Such considerations combine to indicate that an assumption of the structural infinity of the physical extent of the natural universe or of the functional infinity of its nomic or operational complexity is simply not required to provide for the prospect of ongoing scientific progress. Continuing discovery is quite as much a matter of how we inquirers proceed with our work as it is of the nature of the object of inquiry. The salient point is that it is *cognitive* rather than *structural or operational* complexity that is the key here, and that cognitive depth need not depend on physical depth.

7. Lessons

The preceding deliberations underwrite several important philosophical lessons:

(1) The first is that the conclusion we should draw is not *scepticism* (that we cannot obtain any useful knowledge about reality), but rather *fallibilism* – that such information as we can come by is always in some ways imperfect and capable of improvement.

(2) The second is that in our efforts to secure knowledge of the world cannot expect *perfectibility* but rather *improvability*. And that accordingly a certain *epistemic humility* is always in order, seeing that we must acknowledge meliorism – the idea that what we have is imperfect, and that no matter how hard we ourselves may try, the successors who will eventually stand on our shoulders will be able to do better. We can aspire to progress but not to completion.

(3) The third lesson is that of the *untenability of a radical idealism*. We must recognize that reality as such outruns the reach of our own imperfect efforts and that we cannot now (or ever!) identify reality as such with our ever-imperfect picture of it – with reality-as-we-now-think-of-it.

(4) A fourth crucial lesson is that of *pragmatism*. Since we are bound to see our view of the real as defective and imperfect in ways that we cannot monitor by comparing our imperfect picture of nature with reality as such, we have to number the adequacy of our efforts by services that are not theoretical but practical in nature. That is we have to assess the comparative theoretical adequacy of our scientific theories in terms of their capacity to yield successes in the domain of prediction and control over nature.

Beyond these specific lessons there becomes a larger, more ideological conclusion: *Reality* and its associated conception of "the real world" is a somewhat equivocal idea. For we have to do not so much with reality itself as reality – as in such modes as:

- reality as we experience it ("optical illusions" included);
- reality as we *think* that it will be judged to be if and when all returns are in;
- reality as it will *actually* turn out to be once all returns are in.

Our picture of "the real world" encompasses such alternatives as the world-picture of subjective experience, present-day science, perfected science as we estimate it, and perfected science as such. Now seeing that perfected science lies beyond our grasp and that all the returns will never be in we are not in a position to lay any warranted claims to definitiveness and to make unqualifiedly confident assertions about reality as such.

We have to reckon in matters of scientific theorizing with Hegel's cardinal principle that only the totality contains truth – that definitive truth claims require complete evidence and total information, while in actuality our claims to truth at the level of generality and precision at issue in theoretical science are always based on incomplete evidence and partial information. What we deem to be true is always a shadow of the Truth. Our most secure "knowledge" is no more than a best estimate – our scientific successes no more than preparations for further work. In consequence, the very idea of scientific knowledge is itself an idealization. For such inquiry is the pursuit of an unattainable ideal – the ideal of a perfected science that affords us a true and fully adequate account of how things work in the world. We realize full well that this ideal is unachievable. We have no prior assurance on the basis of general principles that the cognitive venture will succeed in significant measure; we can only presume this in the light of our commitment to an ideal, a commitment whose justification comes only ex post facto through its demonstrated utility.

The *epistemic* condition of homo sapiens is much like our *moral* condition. We are destined in both regards to what is at best and most a life of effort and struggle that offers the prospect of achieving not perfection but amelioration. Being a good inquirer, like being a good person, is not a solvable and completable problem but a challenge and an opportunity that affords the prospect of an unending but also a potentially unendingly rewarding quest.[11],[12]

NOTES

1. Transl. by Philip P. Wiener (Princeton, 1954).
2. On this issue see Jesús Mosterín "El mundo se nos escurre entre las mallas de nuestras teorías," *Theorema*, 12 (1982): 181–94.
3. One possible misunderstanding must be blocked at this point. To *learn* about nature, we must *interact* with it. And so, to *determine* a feature of an object, we generally have to make some *impact* upon it that modifies its otherwise obtaining condition. (The indeterminacy principle of quantum mechanics affords a well-known reminder of this.) It should be clear that this matter of physical interaction for data acquisition is not contested in the ontological indifference thesis here at issue.
4. "Remarks by David Bohm," in *Observation and Interpretation*, ed. by Stephan Koerner (New York and London, 1957), p. 56. For a fuller development of Bohm's views on the "qualitative infinity of nature," see his *Causality and Chance in Modern Physics* (London and New York, 1957).
5. A later writer put it this way: "Go on as far as we will, in the subdivision of continuous quantity, yet we never get down to the absolute point. Thus scientific method leads us to the inevitable conception of an infinite series of successive orders of infinitely small quantities. If so, there is nothing impossible in the existence of a myriad universes within the compass of a needle's point, each with its stellar systems, and its suns and planets, in number and variety unlimited." (W. S. Jevons, *The Principles of Science*, 2nd ed., London, 1877, p. 767).
6. The idea of successive of "layers of depth" in the analysis of nature, each giving rise to its own characteristic body of laws in such a way that each of these law-manifolds is more encompassing than its predecessors and that their successive discovery represents a sequentially deeper penetration of the structure of nature was, so far as I know, first mooted in contemporary physics by E. P. Wigner, "The Limits of Science," *Proceedings of the*

American Philosophical Society, 94 (1950): 422–27. But it had already been contemplated by Charles Sanders Peirce, half a century earlier.

7. "The Concept of Probability in the Frame of the Probabilistic and the Causal Interpretation of Quantum Mechanics," in Stephan Koerner (ed.), *Observation and Interpretation* (New York and London, 1957), p. 77. Cf. also E. P. Wigner, *op. cit.*
8. For an interesting and suggestive analysis of "the architecture of complexity" see Herbert A. Simon, *The Sciences of the Artificial* (Cambridge, MA, 1969).
9. John Herschel, *Familiar Lectures on Scientific Subjects* (London, 1867), p. 458.
10. This idea that our knowledge about the world reflects a *interactive* process, to which both the object of knowledge (the world) and the knowing *subject* (the inquiring mind) make essential and ultimately inseparable contributions, is elaborated in several of the author's books, in particular in *Conceptual Idealism* (Oxford, 1973), and *Methodological Pragmatism* (Oxford, 1977).
11. Some of the themes of this chapter are also discussed in the author's *The Limits of Science* (Berkeley and Los Angeles, 1984).
12. An abbreviated version of this essay was presented at the Argentine National University of Córdoba in February of 1993 at the ceremony at which the author was awarded the degree of *Doctor honoris causa*.

CHAPTER 4

METAKNOWLEDGE AND ITS COMPLICATIONS

1. HIGHER-ORDER KNOWLEDGE

The reflexive aspect of human cognition is one of its most characteristic and significant features. Nothing is more significant for and characteristic of our human cognitive situation than our ability to step back from what we deem ourselves to know and take a critically evaluative view of it.

The development of metaknowledge – of information about our knowledge itself – is a crucial component of the cognitive enterprise at large. Metaknowledge is higher-order knowledge regarding the facts that we know (or believe ourselves to know); the object of its concern is our own knowledge (or putative knowledge). The prospect of metaknowledge roots in the reflexivity of thought – the circumstance we can have doubts about our doubts, beliefs about our beliefs, knowledge about our knowledge. The development of metaknowledge is a crucial component of *epistemology*, the cognitive endeavor devoted to the study of knowledge at large. In its pursuit we encounter some very interesting but also disconcerting results, seeing that attention to the actual nature of our knowledge yields some rather paradoxical facts.

2. PROBLEMS OF METAKNOWLEDGE

Let *K* be the manifold of propositions that we can appropriately take ourselves to know: the collection of our appropriately (though not necessarily correctly) staked knowledge claims, specifically including the theoretical claims that we endorse in the domain of the sciences. Thus *K* will include all of the claims which, in our considered judgment, represent something that a reasonable person is entitled – in the epistemic circumstances that currently prevail – to claim as an item of knowledge. It consists of:

- the sum-total of presently available (scientific) knowledge;
- what we endorse nowadays as established knowledge at the level of (scientific) understanding;
- what we take to be known knowledge as to how things work in the world.

K, in sum, is the sum-total of what we accept as established knowledge – the stuff of our scientific manuals and textbooks, the resources of reasoning in the scientific literature.

Given this conception of *K*, it is clear that we can appropriately endorse the rule:

For any *p*: If *p* belongs to *K*, then *p*.

We accept this precisely because its membership in our body of accepted knowledge *K* simply reflects our seeing a claim as true; a claim would not be a *K*-member did we not see it as true.

Unfortunately, however, we know full well that we are sometimes mistaken in what we accept as knowledge, even when every practicable safeguard is supplied. And this is true not only for the ordinary knowledge of everyday life but for our scientific knowledge as well. The following considerations confront us with some of the inescapable facts of (cognitive) life:

Scientific fallibilism. We do (or at any rate should) realize full well that our scientific knowledge of the day contains a great deal of plausible error. We are (or should be) prepared to acknowledge that the scientists of the year 3,000 will think the scientific knowledge of today to be every bit as imperfect and extensively correction-requiring as we ourselves think to be so with respect to the science of 300 years ago. To be sure, we can safely and unproblematically make the conditional prediction that *if* a generalization states a genuine law of nature, *then* the next century's phenomena will conform to it every bit as much as those of the last. But we can never – in the prevailing condition of our information – predict with unalloyed confidence that people will still continue to regard that generalization as a law of nature in the future. Scientific progress brings in its wake not only new facts but also change of mind regarding the old ones; what we have in hand are not really certified "laws of nature" as such, but mere *theories* – that is, *laws as we currently conceive them to be in the prevailing state of the scientific art.*

Cognitive imperfection. We are (or at any rate should be) clearly aware of our own liability to error. We cannot avoid recognizing that any human attempt to state the truth of things will include misstatements as well. We do – and should – recognize that no matter how carefully we propose to implement the idea of membership in our "body of scientific knowledge," in practice some goats will slip in along with the sheep. (The sceptic's route of cognitive nihilism – of accepting absolutely nothing – is the only totally secure way to avoid mistakes in acceptance.)

And so, if any induction whatsoever can safely be drawn from the history of science it is this: that much of what we currently accept as the established knowledge of the day is wrong, and that what we see as our body of *knowledge* encompasses a variety of errors. In fact, there are few inductions *within* science that are more secure than this induction *about* science. Despite the occasional over-enthusiastic pretentions of some scientists to having it exactly right in understanding nature's phenomena – to be in a position to say "the last word" – the fact remains that the realities of the scientific situation pose virtually insurmountable obstacles in this regard.[1] Let us consider why this is so.

3. THE IMPERMANENCE OF PRESENT SCIENCE: WHY THEORIES FAIL

Our most sophisticated predictions regarding the world's operations are based upon the theories of science. But what can we predict regarding such theories themselves? Most safely and securely – failure. Its epistemic instability – its change of mind about nature's *modus operandi* – is a fact of life *about* science that is more firmly established and confirmed than any of the facts *within* science itself. Any judgement we can make about the laws of nature – any model we can contrive regarding how things work in the world – is a matter of theoretical triangulation from the data at our disposal. And we can never have unalloyed confidence in the definitiveness of our data base or in the adequacy of our theoretical exploitation of it; there is clearly good reason why no finite body of observations – however large – can ever settle decisively just what the laws of nature are.

In consequence, our best wrought scientific theories are transitory. They arise, make their way to acceptance – be it gradually or dramatically – and then they come to grief and fall out of favor. If historical experience teaches anything about the future, it teaches that we should expect few (if any) of our scientific theories of today to stand in place by the year 3,000, without amendment, modification, revision, or replacement.

But just why do theories fail? What aspect of reality accounts for this state of affairs? A scientific theory is a structure built to house and attuned to the needs of a certain body of experiences – fitted to the conditions of experimentation, observation, and information processing of a particular technological state of the art. As these conditions change, stresses and strains develop which destabilize the theory structure and lead to its eventual collapse. Our scientific theories (like every other human construction) will ultimately encounter destabilizing conditions. The realities of the situation leave us little option but to be fallibilists about science. Analysis of the theoretical general principles of the case and induction from the history of science both indicate that the theories accepted at one stage of scientific inquiry need not carry over to another but can – and ultimately will – be replaced by something substantially different.

We learn by empirical inquiry about empirical inquiry, and one of the key things we learn is that at no actual stage does science yield a final and unchanging result. All the experience we can muster indicates that there is no justification for regarding our science as more than an inherently imperfect stage within an ongoing development. And we have no responsible alternative to supposing the imperfection of what we take ourselves to know. There is no reason to see the posture of our successors as fundamentally different from our own in this respect.

There emerge certain rather paradoxical facts about our knowledge. For much of what we see and accept as knowledge – and in fact knowledge of the highest quality, viz., scientific knowledge – has to be acknowledged as being little more than yet unmasked error. All too clearly, the cognitive resources at our disposal never enable us to represent reality with full adequacy in the present condition of things – irrespective of what the date on the calendar happens to be.

4. The Reality of Limits

It is important to recognize that operating with the distinction between real and putative knowledge – though unavoidable – is a tricky business. We can apply this distinction retrospectively. ("Yesterday I took myself to know that *p* but I was quite wrong about it.") But we cannot apply it to what is presently before us. ("I know that *p* but deem at open to doubt" is in deep semantical trouble on grounds of simple inconsistency.)

There is, clearly, something very frustrating about this sort of metaknowledge. The self-critical aspect of our metaknowledge – the circumstance that it involves highly general claims about the imperfections of our knowledge – endows it with a paradoxical aspect. This is readily brought to view by considering some further examples.

The preface paradox. The conscientious author of a fact-laden book apologizes in the Preface for the error the book contains. "Several friends have read the MS and helped me to eliminate various errors. But the responsibility for the errors that yet remain is entirely mine." But why not simply correct these mistakes? Alas, one cannot. (If only one knew what they were one would of course correct them.) They are lost in a fog of unknowing.

But yet there is a straightforward logical conflict here. One cannot consistently maintain a collection of contentions distributively and yet also assert that they are not all true collectively:

> p, q, r entails p & q & r which is incompatible with not-(p & q & r)

From the standpoint of logic, the things we claim as true severally and individually must also be claimed as true conjointly. And yet this is not realistic. It just does not reflect – always and everywhere – the realities as we must actually acknowledge them.

The preface paradox situation is indicative of a larger predicament. We know or must presume that (at the synoptic level) our science contains various errors of omission and commission – though we certainly cannot say where and how they arise. And it does not matter how the calendar reads. This state of affairs hold just as much for the science of the future as for that of our own day. Natural science is not only imperfect but imperfectable.[2] And this fact has profound implications for the nature of our "scientific knowledge." Our knowledge as an aggregate of accepted claims is one thing. But our meta-cognitive recognition – indeed knowledge – of the presence of error in the whole collection is another of no lesser significance. Both are facts. And our knowledge here is to all intents and purposes consistent with our meta-knowledge. The paradox is that our metaknowledge *conflict* with our knowledge, seeing that one of the things we cannot avoid adding to our knowledge is the item of metaknowledge that some of our knowledge claims are mistaken.[3]

Blind spots. There are not only errors of comission but errors of omission as well – blind spots as it were. We know *that* there are facts we do not know, though we cannot say *what* these items of ignorance are. We know *that* there are answers we cannot give to some questions, but do not know what those missing answers are. We realize the incompleteness of our knowledge – and can even often localize where this incompleteness exists – but of course cannot characterize those missing items of information positively by identifying them as an individuated item.[4]

5. Unproductiveness of (Some) Metaknowledge

Where knowledge is genuinely deficient we cannot use existing knowledge to fill it in obliquely. We cannot use presently available information to anticipate future discoveries. We can make conjectures about what those future discoveries might be about, but we cannot *ex hypothesi* preview their substance. If we could were to make those future discoveries now they would (*ex hypothesi*) not fall under the rubric of future discoveries but would count as presently available.

The Preface Paradox indicates that we cannot use present knowledge to *correct* itself – even where we recognize and acknowledge that it needs correction. And our deliberations here indicate that we cannot use present knowledge to *complete* itself even though we know it requires completion. Either way, we have the predicament of remaining powerless in the face of acknowledged shortcomings.

The recognition of the cognitive defects which in our metaknowledge regarding our cognitive imperfections reveals to us nevertheless leaves in a position of unavoidable (and unavoidably recognized) impotence. At the end of the day, the recognition of errors of omission or commission in our knowledge does us no good – where such knowledge does not put us in a position to correct such errors. As regards positive, substantive information such defects in our knowledge fall into blind spots.

The irony here is that in such situations metaknowledge proves unavailing. It teaches us about the imperfection in our base-level knowledge in ways that does us little good. Such metaknowledge is informative but nevertheless ineffective. There is, in effect, little or nothing that we can really do about it; it leaves us without help towards removing the deficiencies whose existence it draws to our attention.

6. Productive Metaknowledge: The Recognition of Bias

There are, however, other, more productive cases of metaknowledge, situations in which our recognition of cognitive shortcomings in what we have accepted as "knowledge" does indeed enable us to take steps in the direction of correcting its recognized flaws. This happens in particular in situations of recognized bias.

Predictive biases. People's best-made predictions – like their best-made plans – "gang aft agley." We can safely predict that the cost of any major construction project will exceed our most carefully constructed cost prediction; and we can safely predict that its actual completion will postdate our most carefully contrived prediction of the date of its accomplishment. In this way, we can come to recognize that our most carefully contrived predictions will in certain areas go *systematically* amiss. This sort of thing also happens in other prediction situations. For example, a statistical analysis of weather forecasts shows that meteorologists systematically underestimate the *change* in temperatures from one day to the next by approximately ten percent. Now it is clear that metaknowledge of *this* sort can be eminently productive. For we can indeed make effective use of it to *correct* our errors, proceeding to replace our first-order predictions by second-order predictions designed to correct them in the light of the systematic biases revealed by experience at the first-order level.

Estimation biases. Predictive biases of the just-indicated sort are reflections of the more general situation of estimation biases. For we can come to recognize that in various situations of estimation there is a general tendency of a displacement of people's best estimates of various factors on the direction of too big, too little, too long, too short, etc. And then, of course, we can make straightforward use of such metaknowledge regarding the error tendencies in people's base level knowledge to effect second-order improvements. For example, the combination of estimates drawn from different sources or methods (say the *average* for the sense of simplicity) will generally be more accurate than the best estimate of *most* of the individual contributions.[5]

As examples of this sort show, there is thus also the prospect of *productive* metaknowledge – of taking such apparent knowledge as our predictive and estimative efforts afford to us and *replacing* it by something else that we have good reason to deem him more worthy of recognition as knowledge.

Unfortunately, however, this happy situation is not universal. As we have seen, sometimes metaknowledge can sometimes prove to be merely unproductive but actually troublesome. Let us reexamine this situation and its ramifications more closely.

7. AN APORY AND ITS RECONCILIATION: *K*-DESTABILIZATION

It is helpful to look once more at some individually plausible but collectively inconsistent beliefs about the nature of our knowledge. Let us return to the situation already sketched above regarding our commitment – or seeming commitment – to the (putative) facts constituting our body of knowledge K:

(I) K includes only truths: If $p \in K$, then p

After all, if we did not see p as being true, then we would not have included it in K in the first place. But as we have seen, it is pretty well inescapable that we also stand committed to:

(II) K includes some falsehoods: $(\exists p)\,(p \in K \;\&\; {\sim}p)$

And these two theses are clearly contradictories: they stand no patent conflict with one another. How are we to respond to this situation?

Clearly there is just no way of getting around (II). No induction from the history of science stands more secure than this idea that some of the theories we currently accept as true are just plain false.

But (II) is an item of meta-knowledge that destabilizes (I). And so we have no real choice but to abandon (I) – or rather, to weaken it. After all, we here confront a fundamental axiom of philosophical method: "Never simply abandon. In the face of insuperable difficulties, introduce distinctions and qualifications that enable you to save what you can of your commitments."

Let us inventory the available ways of effecting the unavoidable weakening of thesis (I). The object of the enterprise here, is to make a (hopefully complete) survey of the avenues open to us for obtaining an at least plausible weaker variant of this thesis. The various available possibilities are mapped out in Table 1. Let us briefly examine the tenor and tendency of each of these positions.

1. *Probabilism* is an historic position which was variously espoused in the middle ages as a response to radical scepticism regarding human knowledge in general. (It was already anticipated in antiquity by some of the moderate sceptics of the Middle Academy.) More recently, it has been adopted with respect to scientific theories by the mainstream logical positivists and logical empiricists of the pre-war era, in particular Rudolf Carnap and Hans Reichenbach.

2. *Sophistry* is also an historic position that goes back to classical antiquity. It would have us see the members of K as represent-

TABLE I

A SURVEY OF POSSIBILITIES
(I) If $p \in K$, then p is true POSSIBLE QUALIFICATIONS OF THIS THESIS: Change the ending from "is true" to: A. is (only) *probably* true. (PROBABILISM: what we see as true is no more than probably so.) B. is (only) *putatively* true. (SOPHISTRY: *K* consists of mere opinion.) C. is (only) *putatively* BUT not actually true. (RADICAL SCEPTICISM: Our so-called "knowledge" is no more than currently accepted falsehood.) D. is (only) *approximately* true. (APPROXIMATIONISM: *K*-membership means no more than rough correctness – "being in the right ball-park.") E. is (only) *presumptively* true. (PRESUMPTIONISM or FALLIBILISM: *K* consists of tentative estimates – claims we are only provisionally to endorse as truths, subject to future defeat.) F. is *generally* (or *almost always*) true. (STATISTICISM or REGULISM We change (I) from a universal generalization to a general – but exception-admitting – rule.) G. is true *provided* that p also belongs to $K^{+} \subset K$. (DISCRIMINATIONISM The idea that *K* consists of a secure core (K^{+}) and a more problematic remainder (K^{-}).) H. is *without any claim or pretentions to truth as such*, being merely a practical expedient to guide our conduct in certain regards. (INSTRUMENTALISM) Plan Z. Accept inconsistencies and take them in stride. (INCONSISTENCY TOLERANCE.)

ing no more than plausible opinion. Present-day "postmodernism" to all intents and purposes returns to this position – as does the sociology of knowledge school of the so-called "strong program."

3. *Scepticism* represents perhaps the most drastic step – that of accepting... *nothing at all* and remaining *adoxastos* (doctrinally uncommitted), as the Greek sceptics put it. It calls for simply suspending judgment in matters of scientific theorizing. This posi-

tion has a long and depressing history for classical antiquity to our own day with exponents ranging from Sextus Empiricus to Peter Unger. On this approach, scientific knowledge is seen as effectively amounting to currently accepted error that hasn't been unmasked as such yet. This, in essence, was Nietzsche's view: what we see as our "scientific" knowledge as merely a matter of useful illusion.

4. *Approximationism* or something pretty close to the instinctive views of most scientists, who hesitate to claim that they've got us exactly right, but like to think they've at least got it pretty much right – that our favored scientific theories are correct "in the essentials" as it were. This was – roughly speaking – the view of C. S. Peirce.

5. *Presumptionism or Fallibilism.* On this view, a scientific theory is a temporary expedient to be replaced when and if something better comes along. One commitment to it is merely provisional and tentative – though, to be sure, we can only *displace* something with something else. One could characterize the falsificationist position of Karl Popper as a version of this approach.

6. *Regulism or Statisticism.* This position calls for replacing (I) by "K includes *mainly* truths," and being prepared to tolerate exceptions. Here (I) is read as stating merely a general (but not strictly universal) rule – one which, as such, admits of exceptions, though of course we cannot identify them in advance. In the context of scientific theories, this approach has no historical precedents that I know of.[6]

7. *Discriminationism.* This represents a rather different line of approach and proceeds by qualifying (I) in yet another rather different direction. Its strategy is to divide K into two components: the unproblematic K^+ (whose members must, of course, be mutu-

ally consistent), and the problematic (K^-). And then it proposes to adopt:

(I′) If $p \in K^+$, then p
[Note that the membership of K^+ must accordingly be seen as individually certain and collectively consistent.]

(II″) Not-(If $p \in K^-$, then p) or equivalently
$(\exists p)(p \in K^- \ \&\ \text{not-}p)$

There does not seem to be on record any theoretician who has adopted this discriminationist position, though there is (as we shall see) much to be said for it.

8. *Instrumentalism.* This view of the matter takes science out of the business of claims to truth. Its work is seen not as providing answers to substantive questions but to providing practical guidance for the conduct of affairs in matters of application and of inquiry. Our scientific theories are thus not actual truth-purporting beliefs but belief-neutral instruments for predicting and controlling the phenomenon. This position – or something substantially like it – has been held by some pragmatists (e.g., John Dewey) – and by such 20th century philosophers of science as P. W. Bridgmen and Bas van Fraassen.

9. *Inconsistency Tolerance.* This rather drastic approach of simply taking consistencies in stride view is in favor among some logicians and was to some extent favored by Ludwig Wittgenstein as regards pure mathematics. Some physicists have entertained the idea in with quantum mechanics. However philosophers of science have been distinctly disinclined to resort to so radical a recourse.

It must be noted (and conceded) that these several positions are probably not exhaustive and that for sure they are in many cases not mutually exclusive. Thus, for example, one can combine instrumentalism and discriminationism by relegating all scientific *theories* to the "problematic remainder," while positioning particular observation reports ("These are creatures of the moon") within the secure core. Here one would take the stance that we do not actually *believe* these problematic theories but see them as mere instruments for predicting true core statements.

8. Costs and Benefits

The crucial fact about these various alternative approaches is that none of the options that they afford us are cost-free: each has its characteristic balance of advantages and disadvantages, of costs and benefits. Let us examine somewhat more closely the issue of the costs of adopting the various plans.

Plan A (Probabilism). The domain of the probable is not inferentially closed. So there is now no justification for simply combining a variety of accepted propositions as collective premisses for inference. We arrive at a situation of deductive disfunctionality. Moreover, we cannot now make our way to any definite (unprobabilistic) assertion. We secure no answers to our questions about how things work in the world. (To distribute probabilities across a spectrum of alternatives is emphatically not to provide an answer.)

Plan B (Sophistry). In its substance, this is a know-nothing position, that destroys any prospect of a workable distinction between knowledge and opinion.

Plan C (Radical Scepticism). This position is simply a cognitive nihilism. By seeing all of our scientific knowledge as categorically mistaken it throw out the baby with the bath water. It is an all-annihilating H-bomb that allows us no answers, no guidance, no nothing.

Plan D (Approximationism). This approach confronts us with the notorious difficulties of defining approximation. If we claim to have it right "in its essentials," it would be nice to know (as we do not and cannot) what those essentials are. Then too there are all those problems of confirming approximations. (In cognitive contexts, is somewhere between difficult and impossible to pre-establish how matters will evolve over the long run.)

Plan E (Presumptionism, Fallibilism). On this basis we can only stake only guarded claims to truth. We insist that whatever we maintain as true need hold only *pro tem*, to be replaced when and if something better comes along. (Note that we must here differentiate between "is presumably true" and "is presumably true.")

Plan F (Regulaism/Statisticism). This position indulges a certain over-optimism. For it does not do justice to the extent to which K has problems, seeing that K is simply not plausible to see the bulk our present scientific theories as exact and definitive truths.

Plan G (Discriminationism). This tactic confronts us with the burden of delineating that unproblematic case $K^+ \subset K$. It saddles us with the problem of differentiating K^- from K^+ – of identifying (at least some of) unproblematic knowledge.[8]

Plan H (Instrumentalism). This has the disadvantage of cognitive vacuity. Like scepticism (to which it is closely akin) it leaves us without answers to our questions.

Plan Z (Inconsistency Tolerance). This approach would have us simply take the inconsistency between principles (I) and (II) in stride, and to accept at face value the incoherence of what we know. This involves abandoning the traditional idea that it is a non-negotiable demand of rationality to insist on the consistency of one's knowledge claims. The modern semantico-logical theory of inconsistency tolerance ("paraconsistency") has created mechanisms for implementing this seemingly absurd strategy in a more or less workable way.[9] This approach represents so radical a recourse, however – one that cuts so deep a swath across so large a terrain – that it cannot count as more than a plan of last resort, a distantly desperate expedient to be adopted if and when all else fails.

9. FINAL RECKONING

On the basis of these considerations, it would seem to emerge that what is to all appearances the best available course of resolution is that afforded by combining a presumptivistic Fallibilism (Plan E) with a judicious Discriminationism (Plan G). In effect, this strategy calls for endorsing two theses:

(1) That we see the scientific theories of the day as representing – in the main – not categorical truths, but rather merely *best estimates* that we can accept provisionally and *pro tem*, as viable truth surrogates adopted with due tentativity until such time as something better comes along;

(2) That while this position is something which we adopt on a "by and large" and "in general" basis (symbolized by that qualification "in the main"), we nevertheless stand prepared to endorse as categorically true certain particular categories of contentions, so that we do not see *everything* as fallible. On the one hand this secure core (K^+) of categorically endorsed contentions will include such predominantly *observational* information as "There are craters on the moon." On the other hand, as regards predominantly *theoretical* information, this unproblematic core will also contain (among others) two further groups:
   1. *Vague knowledge* (schoolbook science). [For example, "The structure of matter is atomic."][10];
   2. *Metaknowledge*. [For example, "The bulk of our scientific theories will have to be replaced or emended."][11]

This position may be motivated on the basis of a cousin of the epistemological "Sherlock Holmes Principle" : "When you have eliminated the impossible, then whatever remains, however improbable, must be the truth." (A. Conan Doyle, "The Sign of Four" (1890), ch. 6.) However, in the present context, this precept should be so transmuted so as to take variant form of an *axiological* principle: "When you have eliminated the even less appealing alternatives, then whatever remains, however unappealing, must be accepted." That is – when all the other alternatives are even less appealing, then a given alternative, however unappealing, must be endorsed. (In effect, this represents an axiologically geared policy of rational choice in the context of a fixed range of alternatives.)

It is on this basis of comparative cost-effectiveness (rather than any claims to inherent and absolute appropriateness) that these deliberations urge the claims of a Fallibilistic Discriminationism – or, perhaps better, a Discriminating Fallibilism – upon the reader's sympathetic consideration.

NOTES

1. On this issue see the author's *Scientific Progress* (Oxford: Basil Blackwell, 1978). As the discussion there makes clear, what is at issue is not *just* an induction from the history of science; there are also reasons of fundamental general principle why the scientific investigations of the future are pretty much foreordained to destabilize the findings that result from the investigations of the past.
2. For relevant considerations see also the author's *The Limits of Science* (Berkeley and Los Angeles, 1985).
3. On the Preface Paradox see A. N. Prior, "On a Family of Paradoxes," *Notre Dame Journal of Formal Logic*, 2 (1961): 26–32; D. C. Makinson, "The Paradox of the Preface," *Analysis*, 25 (1965): 205–7.
4. On these and related issues see Roy A. Sorensen, *Blindspots* (Oxford, 1988).
5. For substantiation see R.G. Winkler and S. Makridakis, *Journal of the Royal Statistical Society*, 146 (1983): 150–57.
6. See however the author's *Philosophical Standardism* (Pittsburgh, 1994) for the elaboration of a version of this approach.
7. For example in the work of Richard Sylvan. And see also Nicholas Rescher and Robert Brandom, *The Logic of Inconsistency* (Oxford, 1979).
8. Effectively subject to the rule that when there is a conflict the members of K^- must give way to those of K^+. On those matters see the author's *Plausible Reasoning* (Assen, Amsterdam, 1976).
9. See, for example, N. Rescher and R. Brandom *The Logic of Inconsistency* (Oxford, 1979).

10. On this issue see the author's *Scientific Realism* (Dordrecht, 1987).
11. For further details see the author's *The Limits of Science* (Berkeley and Los Angeles, 1984).

CHAPTER 5

FALLIBILISM AND THE PURSUIT OF TRUTH

1. Fallibilism

Fallibilism is a philosophical doctrine most closely associated with the work of C. S. Peirce.[1] The doctrine is best understood in terms of a contrast with some related positions regarding the prospects of scientific knowledge of the world about us:

- all claims to theoretical scientific knowledge are inappropriate: We humans cannot achieve real knowledge in this domain, none of our claims to knowledge here can ever be satisfactorily substantiated. (SCEPTICISM);
- what we claim as theoretical scientific knowledge is always inaccurate and approximate – sufficient by and large for our practical purposes, but never accurate and strictly speaking correct. (APPROXIMATIONISM);
- our theoretical scientific knowledge claims are always vulnerable: they must always be staked tentatively because the prospect that further inquiry and discovery will lead to their modification or replacement can never be eliminated; the things we take ourselves to know many in the end have to be abandoned. (FALLIBILISM).

Fallibilism is tantamount to a probabilism which holds that we can never stake a warranted claim to secure knowledge; all our knowledge claims in theoretical science possess varying degrees of probability and what they assert should never be seen as certain. For if each and every one of our knowledge claims here is vulnerable and may turn out to be false, then none can be asserted as true categorically, but can only be maintained as having some probability of turning out to be true. And conversely, if all our knowledge in theoretical science is merely probable, then any particular claim is vulnerable and may turn out to be false.

The following further points regarding the interrelationship of the three indicated doctrines deserve particular note: (1) Both of the other doctrines are at odds with theory scepticism. Approximationism does not reject all such knowledge claims as altogether inappropriate but sees them as falling into the general neighborhood of the truth. Fallibilism, however sees these claims as justifiable and indeed of having some chance of getting it right. (2) Approximationism regards all of our knowledge claims in theoretical science not as probably true but as always certainly false as they stand – that is, each such claim asserts something that is not literally true as formulated but merely falls somewhere in the general neighborhood of the truth. So strict approximationism is incompatible with a fallibilism that sees our truth claims as possibly true though probably false. (3) However, approximationism can be relaxed and combined with fallibilism in a doctrine which maintains (i) that it is not only possible but virtually certain that our knowledge claims in theoretical science will be defeated and will have to give way to others, and (ii) that these other, later (and presumably superior) knowledge claims manifest their superiority by being *better approximations* to the actual truth of the matter. With respect to natural science, it is exactly this last-indicated doctrine of a combined fallibilism-cum-approximationism that characterized the epistemic theory of Charles Sanders Peirce. Other recent thinkers who have espoused fallibilism with respect to the theories of natural science include Gaston Bachelard, Rudolf Carnap, Pierre Duhem, and Karl Popper. The mitigated sceptics of the middle academy were the precursors of this position, since they taught that we can never certain knowledge (*epistêmê*) in theoretical matters but have to make do with what is nearly probable or plausible (*to pithanon*).

The alternatives to fallibilism/probabilism have deep problems. Radical scepticism leaves us empty-handed, without any answers to our questions. Approximationism faces the difficulty that it is problematic to claim to approximate something itself forever inaccessible. Yet what is it that speaks for fallibilism in theoretical science? Why should we see all of our claims to knowledge of the world – even our best state-of-the-art scientific theories – as being vulnerable?

As fallibilism sees the matter, we have no assurance that our scientific theories or systems are definitely true; they are simply the best we can do here and now to resolve our question regarding nature's *modus operandi*. Any such scientific theory or system is the product of human contrivance. And any such contrivance – be it a house, a dam, or a knowledge claim – is vulnerable. The processes of change that come with time always involve chance eventuations that bring new, unforeseen and unforeseeable circumstances to the fore. Every structure, be it physical or intellectual, is thus ultimately likely to encounter conditions that its constructors did not – and could have – foreseen. And this circumstance renders its ultimate failure likely. Changed social conditions destabilize social systems; changed physical conditions, destabilize physical structures; changed experiential (i.e., experimental and observational) conditions – changed scientific technology, if you will – destabilize scientific theories. Rational inquiry links the products our understanding to the experienced conditions of a world in which chance and chaos plays an ineliminable role, so that there will always be new relations that ultimately threaten our rational contrivances. (Of course, while we can safely predict *that* our scientific theories will fail – will have to be replaced or modified – we cannot foresee *how* these replacements and modifications will be configured.)

There is much in this picture of the cognitive situation that rings true. The fact is that the equilibrium achieved by natural science at *any* given stage of its development is always an unstable one. The subject's history indicates that scientific theories have a finite lifespan; they come to be modified or replaced under various innovative pressures, in particular the enhancement of observational and experimental evidence (through improved techniques of experimentation, more powerful means of observation and detection, superior procedures for data processing, etc.). As C. S. Peirce and Karl Popper after him have insisted, we must acknowledge an inability to attain the final and definitive truth in the theoretical concerns of natural science – in particular at the level of theoretical physics. Our present science cannot plausibly claim to deliver a definitive picture of physical reality. We would like to think of our science as "money in the bank" – as something safe, solid, and reliable. Unfortunately, however, history militates against this comfortable view of our scientific theorizing. In science, new knowledge does not just supplement but generally upsets our knowledge-in-hand. We must come to terms with the fact that – at any rate, at the scientific level of generality and precision – *each* of our accepted beliefs *may* turn out to be false, and *many* of our accepted beliefs *will* turn out to be false. After all, error is infectious. If you reason from various premisses only one of which is incorrect, the odds are that your conclusion will also be incorrect.

We learn by empirical inquiry about empirical inquiry, and one of the key things we learn is that at no actual stage does theoretical science yield a final and unchanging result. We have no responsible alternative to supposing the imperfection of what we take ourselves to know. (And there is no reason to see the posture of our successors as fundamentally different from our own in this respect.) We occupy the predicament of the "Preface Paradox" exemplified by the author who apologizes in his preface for those errors that have doubtless made their way into his work, and yet blithely remains committed to all those assertions he makes in the body of the work itself. We know or must presume that (at the synoptic level) there are errors, though we certainly cannot say where and how they arise.

To protect scientific realism in the face of the prospect of unending change in the fabric of scientific theorizing, it is tempting to adopt Charles Sanders Peirce's strategem of a resort to *convergent approximation*. This calls for envisaging a situation where, with the passage of time, the theories we successively arrive at grow increasingly concordant and their claims less and less differentiated. In the face of such a course of successive changes of ever-diminishing significance, we could proceed to maintain that the world really is not as *present* science claims it to be, but rather is as the ever more clearly emerging science-in-the-limit claims it to be. The reality of ongoing change is now irrelevant as with the passage of time the changes matter less and less. We increasingly approximate an essentially stable picture.

This prospect is certainly a theoretically possible one. But neither historical experience nor considerations of general principle provide reason to think that it is actually possible. Indeed, quite the reverse!

Any theory of convergence in science, however carefully crafted, will shatter against the *conceptual innovation* that continually brings entirely new, radically different scientific concepts to the fore, carrying in its wake an ongoing wholesale revision of "established fact." Investigators of an earlier era not only did not *know* what the half-life of californium was, but they would not have *understood* it even if this fact had been explained to them. The questions we can pose are limited by our conceptual horizons. It did not appear to be a realistic prospect to all those late nineteenth-century physicists who investigated the properties of the luminiferous aether that no such medium for the transmission of light and electromagnetism might exist at all. Ongoing scientific progress is not simply a matter of increasing accuracy by extending the numbers in our otherwise stable descriptions of nature out to a few more decimal places. Significant scientific progress is genuinely revolutionary in involving a *fundamental change of mind* about how things happen in the world. Progress of this caliber is generally a matter not of adding further facts – on the order of filling in a crossword puzzle – but of changing the framework itself. And this circumstance blocks the theory of convergence.

In any convergent process, later is lesser. But since scientific progress in matters of fundamental importance is generally a matter of replacement rather than mere supplementation, there is no reason to see the *later* issues of science as *lesser* issues in the significance of their bearing upon science as a cognitive enterprise – to think that nature will be cooperative in always yielding its most important secrets early on and reserving nothing but the relatively insignificant for later on. The fact is that a very small scale effect – even one that lies very far out along the extremes of a "range exploration" in terms of temperature, pressure, velocity, or the like – can force a far-reaching revolution and have a profound impact by way of major theoretical revisions. (Think of special relativity in relation to aether-drift experimentation, or general relativity in relation to the perihelion of Mercury.) Given that natural science progresses mainly by substitutions and replacements that go back to first principles and lead to comprehensive overall revisions of our picture of the phenomena at issue, it seems only sensible to say that the shifts across successive scientific "revolutions" maintain the same level of overall significance – that scientific revisionism is neither a convergent nor a divergent process.

With any sort of estimate, there is always a characteristic tradeoff relationship between the evidential *security* of the estimate on the one hand (as determinable on the basis of its probability or degree of acceptability), and its contentual *definiteness* (exactness, detail, precision, etc.) on the other. But technical science forswears the looseness of vague generality or analogy or approximation. It has no use for qualifiers such as "usually," "normally," or "typically;" universality and exactness are its touchstones – it declares not merely that roughly such-and-such generally occurs in certain sorts of circumstances but exactly what happens in exactly what circumstances. The law claims of science involve no hedging, no fuzziness, no incompleteness, and no exceptions; they are strict: precise, wholly explicit, exceptionless, and unshaded. In making the scientific assertion "The melting point of lead is 327.5°C," we mean to assert that *all* pieces of (pure) lead will unfailingly melt at *exactly* this temperature. We certainly do not mean to assert that *most* pieces of (pure) lead will probably melt at *somewhere around* this temperature. In science we always aim at the maximum of universality, precision, accuracy, and exactness. And this commitment to generality and detailed precision renders the claims of science highly vulnerable. We realize that none of the hard claims of present-day frontier natural science will move down the corridors of time untouched. Fragility is the price that we pay in science for the sake of generality and precision.

The interesting fact is that fallibilism is a more plausible doctrine with respect to *scientific* knowledge than with respect to the less demanding "knowledge" of everyday life. Ordinary-life generalizations are such as to allow one who asserts (say) that peaches are delicious, to be asserting something like, "Most people will find the eating of suitably grown and duly matured peaches a rather pleasurable experience." Such a statement has all sorts of implied safeguards, such as "more or less," "in ordinary circumstance," "by and large," "normally," "if all things are equal," and so on. They are thus so well hedged that it is eminently implausible that contentions such as these should be overthrown.

In science, however, we willingly accept greater cognitive risks because we ask much more of the project. Here the objectives are primarily theoretical and governed by the aims of disinterested inquiry. Accordingly, the claims of informativeness – of generality, exactness, and precision – are paramount. We deliberately court risk by aiming at maximal definiteness and thus at maximal informativeness and testability. Aristotle's view that terrestrial science deals with what happens ordinarily and in the normal course of things has long ago been left by the wayside. The theories of modern natural science have little interest in what happens generally or by and large; they seek to transact their explanatory business in terms of strict universality – in terms of what happens always and everywhere and in all kinds of circumstances. We therefore have little choice but to acknowledge the vulnerability of our scientific statements, subject to the operation of the security-definiteness trade-off. Ironically, then, the "common sense" of everyday life is securer than the "well-established knowledge" science.

The ineliminable prospect of error – the recognition of the potential corrigibility of all our scientific theorizing, and the unavoidable acknowledgement that we are not in a position to claim that in this domain *our* truth is *the* truth – mean that we cannot ever lay claim to a definitively correct and final picture of reality. Once we acknowledge that a prospect of incompleteness and a presumption of at least partial incorrectness attaches to our present picture of the world, we can no longer subscribe to the idea that the world really exists as we conceive of it. And so we can no longer adopt the view that, at the level of scientific theorizing, "our world picture" depicts "the real world" – the world as it actually is.

The fact is that we have no choice but to view what we do get hold of by way of "scientific truth" about the world – now, or ever – as being (at the synoptic, systemic level) *imperfect*: incomplete, incorrect, perhaps even inconsistent. The gulf between Appearance (= our *putative* reality) and Reality as portrayed by a body of capital-T Truth that satisfies all of the traditional ideals of completeness, consistency, finality, etc., is – and ever shall remain – every bit as wide as F. H. Bradley envisaged it to be.

Throughout scientific inquiry, the prospect of a gap between "what *we think* the real to be" and "what reality is *actually like*" is absolute and insuperable: we are never in a position to claim it to be closed. It is clear that we have – and can have – no cognitive access to reality except through the epistemic route that our mechanisms of inquiry afford us. Our scientific endeavors provide us no way to determine what reality is like save through the fallible epistemic process of deciding which theories about it we are to accept. We have no way of coming to cognitive grips with reality apart from the laborious, and inherently error-prone, empirical and inductive process of constructing our "picture of the world." The *character* of the world is no doubt independent of our cognitive devisings, but its *characterization* emphatically is not, but is something supervenient and description-relative. And *for us* – and this is the key point – there is no cognitive access to character save via characterization. There is no royal road to knowledge of reality save through the mediation of imperfect inquiry.

Accordingly, we do well to temper our claims to scientific knowledge with a Cognitive Copernicanism. The original Copernican revolution made the point that there is nothing *ontologically* privileged about our own position in space. The doctrine now at issue effectively holds that there is nothing *cognitively* privileged about our own position in time. A kind of intellectual humility is in order – a diffidence that abstains from the hubris of pretentions to cognitive finality or centrality. Such a position calls for the humbling view that just as we think our predecessors of a hundred years had a fundamentally inadequate grasp on the furniture of the world, so our successors of a hundred years hence will almost certainly take the same view of *our* purported knowledge of things. The facts of epistemic life require us to recognize that, as concerns our scientific understanding of the world, our most secure knowledge is very likely no more than presently acceptable error.

2. The Pursuit of Truth

Given fallibilistic insistence on the fragility and inadequacy of human knowledge one may well question the very idea of the "pursuit of truth." Considering that fallibilism has it that the definitive truth lies beyond our grasp, what actual work is the conception of *the truth* able to do for us? Should we perhaps give up on truth?

By no means! To begin with, we indispensably require the notion of truth to operate the classical concept of correct assertion as "agreement with reality" (*adaequatio ad rem*). Once we abandon the concept of truth, then we would also have to abandon the idea that in accepting a factual claim we accept it as true, and thereby become committed to how matters actually stand ("how it really is"). The very semantics of our discourse constrains this coordination of truth and reality: we have no alternative but to regard as real those states of affairs claimed by the contentions we are prepared to accept as true. Once we put a contention forward by way of serious assertion, we must view their claim we accept as representing, at the very least, our best *estimate of the truth*. We thus need the notion of truth to operate the conception of reality. By virtue of their nature as truths, true statements must state facts: they state what really is so, which is exactly what it is to characterize reality. The conceptions of *truth* and of *reality* come together in the notion of *adaequatio ad rem* – the venerable principle that to speak truly is to say how matters stand in reality, in that things actually are as we have said them to be. That is just how these concepts of truth and adequacy work, and they would not work out without that fundamental precommitment to the idea of truth. The very conception of inquiry as we conceive of it would have to be abandoned if the contrast conceptions of "actual reality" and "the real truth" were no longer available.

It thus follows that, in the second place, the nihilistic denial that there is such a thing as truth would destroy once and for all the crucial Parmenidean divide between appearance and reality. And this would exact a fearful price from us: we would be reduced to talking only of what we (I, you, many of us) *think* to be so. The crucial contrast notion of the *real* truth would no longer be available: we would be able to contrast only our *putative* truths with those of others but could no longer operate the classical distinction between the putative and the actual, between what people merely *think* to be so and what actually *is* so. We could not take the stance that, as the Aristotelian commentator Themistius put it, "that which exists does not conform to various opinions, but rather the correct opinions conform to that which exists."[2]

The third point is the issue of cognitive coordination. Communication and inquiry, as we actually carry them on, are predicated on the fundamental idea of a real world of objective things, existing and functioning "in themselves," without specific dependence on us and so equally accessible to others. Intersubjectively valid communication has to be based on the presupposition of a common attempt to get at the truth of things at least as much as the nature of our circumstances permits. All our ventures at communication and communal inquiry are predicated on the stance that we communally inhabit a shared world of things available for contemplation independently of any single person's ideas about them. Where we abandon the commitment to a commonly accessible truth in favor of our varying conceptualization – where we claim for ourselves the exclusive rights to conceptualization (where I conceive of my nightmare and you conceive of yours) – the question of reidentification (of whether your dream matter is the same one as mine) cannot arise, and so the prospect of a sharing of experience about them as such (i.e., as monsters rather than people's ideas about monsters) becomes impossible. It is, to be sure, perfectly possible for two people to communicate effectively about something that is wholly nonexistent and about which they have substantially discordant conceptions (for example, *X*'s putative wife, where *X* is, in fact, unmarried, though one party is under the misimpression that *X* is married to *A* and the other that *X* is married to *B*). Given a shared commitment to an objective framework of things, such aberrations are possible. But this possibility itself attests to our standard commitment to a commonality of communicative focus as a basis on which alone the exchange of information (or misinformation) and the discovery of error becomes possible. The shared (conventionalized) intention to tell the impersonal truth about the same thing undergirds and sustains all our communicative ventures.

This points to a fourth important consideration. Only through reference to the real world as a *common object* and shared focus of our diverse and imperfect epistemic strivings for the truth are we able to effect communicative contact with one another. Our standard view of inquiry takes the line that the real nature of the world is in the main independent of the process of inquiry, one, however, that the real world canalizes or conditions. Dependency is a one-way street here: reality shapes or influences inquiry in its concern for truth, but not conversely. The truth is what we at least try to get at – however gravely our variable individual efforts may fall short. Our knowledge of the world must be presumed incomplete, incorrect, and imperfect, with the consequence that "our conception of the truth" must be considered to afford an inadequate characterization of "the actual truth itself." With respect to our cognitive endeavors, "man proposes and nature disposes," and it does so in both senses of the term: it disposes *over* our current view of reality, and it will doubtless eventually dispose *of* it as well. We must simply do what we can with the relatively feeble means at our disposal to estimate (as best we can) the actual truth of things. And we need the conception of "the real truth" to make sense of the inquiry-pervasive contrast between what we have and what we would ideally like to obtain.

The fifth consideration is that the very idea of inquiry as we standardly conceive it would have to be abandoned if the conceptions of "actual reality" and "the real truth" were no longer available to serve their crucial contrasting roles. We could no longer assert, "What we have there is good enough as far as it goes, but it is presumably not 'the whole real truth' of the matter." Without the conceptions of truth and reality we could not think of our knowledge in the fallibilistic mode we actually use – as having provisional tentative, improvable features that constitute a crucial part of the conceptual scheme within whose orbit we operate our concept of inquiry. For our commitment to the pursuit of "the actual truth" stands together with our acknowledgement that, in principle, any or all of our present scientific ideas as to how things work in the world, at *any* present, may well prove to be untenable. Our conviction in a reality that lies beyond our imperfect understanding of it (in all the various senses of "lying beyond") roots in our sense of the imperfections of our scientific world-picture – its tentativity and potential fallibility. In abandoning our commitment to an objective truth, we would lose the impetus of inquiry.

Inquiry, as we standardly conceive it, is predicated on the commitment to an inquirer-independent truth of things; it is a quest for information about "the real world" with respect to which our own conceptions of things are nowise definitive, and into which others can accordingly enter unproblematically. It is thus geared to the conception of an objective world: a communally shared realm of things that exist strictly "on their own," composing an enduring and impersonal realm within which – and, more important, with reference to which – inquiry proceeds. We could not operate the notion of inquiry as aimed at estimating the true character of the real if we were not prepared to presume or postulate a real truth for these estimates to be estimates of. It would clearly be pointless to project our estimates of the actual truth if we did not stand committed to the idea that there indeed is an actual truth to be estimated. The pivotal contrast in this regard is between "mere appearance" and "reality as such," between "our picture of reality" and "reality itself," between what actually is and what we merely think (believe, suppose, etc.) to be. And our allegiance to the conception of reality and to this contrast that pivots upon it root in the fallibilistic recognition that at the level of the detailed specifics of scientific theory, anything we presently hold to be the case may well turn out otherwise – indeed, certainly will do so if past experience gives any auguries for the future.

In summary, then, we need the postulate of an objective manifold of actual truth for at least five important reasons.

1. To preserve the distinction between true and false with respect to factual matters and to operate the idea of truth as agreement with reality;
2. To preserve the distinction between real and seeming truth, between appearance and reality, between our picture of reality and reality itself;
3. To provide a conceptual basis for intersubjective communication;
4. To furnish the basis for a shared project of communal inquiry;
5. To provide for the fallibilistic view of human knowledge.

Our commitment to the conception of "the actual truth" accordingly plays a central and indispensable role in our thinking with respect to matters of language and cognition. In communication and inquiry alike we seek to offer answers to our questions about how matters actually stand. Our concern is with "the real world." It is seen as the epistemological *object* of veridical cognition, in the context of the contrast between "the real" and its "merely phenomenal" appearances. Again, it is seen as the target or telos of the truth-estimation process at issue in inquiry, providing for a common focus in communication and communal inquiry. (The real world thus constitutes the "object" of our cognitive endeavors in both senses of this term – the objective at which they are directed and the purpose for which they are exerted.) Truth and reality serve us as coordinated conceptions. And these facets of the concept of truth are integrated and unified in the classical doctrine of truth as it corresponds to fact (*adaequatio ad rem*), a doctrine that makes sense only in the setting of a commitment to an inquirer independent reality and an actual truth about it. We stand committed to the idea that there *is* a truth of the matter even in circumstances where we do not *know* it. And indeed we could not acknowledge the insufficiencies of what we purport did we not have the idea of "the real truth" available as a contrast conception.[3]

Some philosophers (Peirce included) see fallibilism as having ethical implications. They project an ethics of belief according to which we have *no right* to claim definitive truth for our current scientific claims, a view which they coordinate with a purported duty for the community of inquirers to pursue inquiry to the greatest extent possible in the circumstances. Accordingly, they insist that the fallibilism that characterizes our scientific endeavors must emphatically *not* be construed as an open invitation to a sceptical abandonment of the scientific enterprise. Rather, it is an incentive to do the very best we can. In human inquiry, the cognitive ideal is correlative with the striving for definitive systematization. And this is an ideal that, like other ideals, is worthy of pursuit despite the fact that we must recognize that its full attainment lies beyond our grasp.

NOTES

1. Charles Sanders Peirce, "Collected Papers, 8 vols. (Cambridge, MA, 1931–58); *Writings of Charles S. Peirce* (Bloomington, 1982ff). For a good general account of Peirce's thought see Murray G. Murphey, *The Development of Peirce's Philosophy* (Cambridge, MA, 1961).
2. Maimonides, *The Guide for the Perplexed*, bk., 1, chap. 71, sec. 96a.
3. Some other aspects of this issue will be considered in Sect. 9 of Chapter 7 below.

CHAPTER 6

METHODOLOGICAL OPTIMISM

1. INTRODUCTION

Since the days of Descartes in the seventeenth century, the dominant tendency in the theory of knowledge has been to accentuate the negative. The threats and temptations of nihilistic doubt and sceptical negativism have been at the center of philosophical concern. The problem-focus has been set by such questions as: "How can one sensibly accept anything at all?" "How can one ever be warrantedly confident of not going wrong in one's information-seeking efforts?" Cognitive pessimism has come to cast its frigid shadow across the entire epistemological scene: even (nay, particularly) dedicated cognitivists saw scepticism as the prime focus of concern.

There is, however, something very problematic and indeed deeply misguided about all this. For it is clear that this negativism is not only opposed to the approach of the older rationalistic mainstream of epistemological tradition in Western philosophy, but that it represents a policy that is neither realistic in relation to the actualities of our epistemological situation nor profitable vis-à-vis our best interests in the business of rational inquiry.

2. PRESUMPTION AND BURDEN OF PROOF

In the epistemic tradition of the West, probative argumentation in reasoned controversy and deliberation has from antiquity onwards centered around the concepts of presumption and burden of proof. These concepts represent ideas that are illustrated by an approach deeply entrenched in classical antiquity and canonized in the approach of Roman law to the idea of burden of proof (*onus probandi*), which have the line that such a burden always rests with the side that actively maintains something (*agents incumbent probatio*), so that the requirement of substantiating one's contentions lies with those who maintain a positive thesis as against those who maintain a negative position (*necessitas probandi incumbit ei qui dicit non ei qui negat*: "The need for proof lies with one who affirms not with one who denies"). While this approach obviously involves concessions to the partisans of negativism, it equally clearly does *not* take a negative and nihilistic line. For, of course – and this point is crucial – any context of rational inquiry *the adducing of supporting considerations must be something that is not made in principle impossible*. Obviously, if any and every contention whatsoever even to meet with rejection if there were an automatic across-the-board presumption of falsity – then it would become in principle impossible ever to provide a persuasive case. Any supportive consideration would automatically be set aside. It is thus clear that the rule that each and every contention needs evidential support through the adducing of further substantiating contentions cannot reasonably be made operative *ad indefinitum*. It makes sense to speak of a "burden of proof" only in the context of established rules regarding the discharge of such a burden. In rational controversy, there must always be some impartially fixed and agreed common ground to determine something that can legitimately be counted as evidence. Not *everything* can be disqualified automatically as failing to qualify in this regard. This line of consideration brings us to the topic of *presumption*.

If adducing evidence in the course of rational argumentation is to be possible at all – if there is to be any sense whatever to the project of rational inquiry – then the idea that any maintained thesis invites an automatic negative response must have its limits. Some claims, at least, must be "innocent until proven guilty." And here we come to presumptions. A presumption is a proposition that has a certain initial plausibility or credibility. It is a contention that is such that, in the absence of specific counter-indications, we are entitled to accept the claim at issue at least "until further notice." With such a defeasible presumption we have what is, in effect, just the reverse side of a burden of proof (of the "burden of further reply" variety). The claim at issue will stand until somehow counter-evidentiated – until displaced and replaced by some positive datum to the contrary. A presumption is thus a claim that stands secure until and unless such a point is reached where something else dislodges it from place. Archbishop Whately long ago formulated the relationship at issue in the following terms:

> According to the most current use of the term, a "presumption" in favour of any supposition means, not (as has been sometimes erroneously imagined) a preponderance of probability in its favour, but such a *preoccupation* of the ground as implies that it must stand good till some sufficient reason is adduced against it; in short, that the *burden of proof* lies on the side of him who would dispute it.[1]

It is in just this sense, for example, that the "presumption of innocence" standing in favor of the accused is correlative with the burden of proof carried by the state in establishing his guilt. Such a *presumptio juris* is a supposition relative to the known facts which, by legal prescription, is taken to stand until refuted. And analogously, there is, in most probative contexts, a standing presumption in favor of the usual, normal, customary course of things.

Of course in any situation where presumption enters in we have also to be concerned about counter-evidence. For evidence can overturn presumptions. But until such time as the available evidence dislodges the presumptively favored alternative it stands in place. The tenability of a presumption is usually tentative and provisional, not absolute and final. A presumption only stands until the crucial issues that "remain to be seen" have been clarified, so that it is actually apparent whether the presumptive truth will in fact remain in place once everything is said and done. In "accepting" a thesis as presumptively true one concedes it a probative status that is strictly provisional and *pro tem*; one does *not* say to it: "Others abide our question, thou art free."[2] There is no need here to invoke the idea of *unquestionable* theses, theses that are inherently uncontestable, certain, and irrefutable.

A presumption as such is thus pre-evidential – at any rate at the level of case-specific particularities. Evidence is something that comes along later on to test – and thereby conceivably defeat – our presumptions. It can (and frequently does) invalidate them and put something else in their place. But, nevertheless, the pivotal idea is that something is already in place before confirmatory or disconformatory evidence comes along. It is a thesis that is tenable alright, but only at the level of "initial credibility."

3. Methodological Optimism and the Presumption of Positivity

We now come to our central theme and thesis in this context. At the core of methodological positivism lies the idea of a certain characteristic – and productively optimistic – approach to the business of inquiry. This approach is based on the *presumption of positivity* that pivots on adopting the stance that in rational problem solving a presumption stands on the side of the most positive alternative – taking the epistemically optimistic line that whatever alternative is the most useful and productive for problem solving will for that very reason be maximally deserving of our favor.

Such a policy of positivity presumption stands correlative with the methodological precept of trying the most promising alternatives first. It is predicated on the idea that where other things are equal we do well to favor the most beneficial alternatives. It turns on the following approach: When other things are everything like equal, opt for the problem solution that is the most promising – that offers the greatest potential benefit in the context of present purpose. Its governing principle is:

Other things equal (or vacuous) the benefit of presumption favor the most beneficial (positive) alternative. And correspondingly the burden of proof inclines against those alternatives that are beneficially suboptimal.

Thus the overall the procedure is one of giving probative weight to positive value (utility) in the relevant context of cognitive operation. On this approach, we have a policy of letting *axiological* – that is, value-geared – consideration be our guide in the context of epistemic problem solving. The presumption at issue does not stand unsupported *in vacuo,* but is indeed well spoken for – if only by considerations of utility.

And the relevant forms of benefit at issue here encompass preeminently the following factors:

- informativeness;
- cognitively utility (productivity);
- cognitively fertility (agenda-enhancing potential);
- simplicity;
- systematicity.

The methodological optimism at issue is thus predicated on a distinct methodological attitude: Feel free to make free use of the epistemic resources you have in hand until something discernibly better comes along. To be sure, you must not be pig-headed and hold on to your position-in-place what may. By all means, be prepared to trade your cognitive instrumentalities in when something better comes along. But not until then. An attitude of open-minded conservatism is called for, and a hopeful view of the prospects of adequately reaching our informative decision problems.

Presumptions, after all, do not fixedly prejudge anything – they are defeasible and defeatable. But they do engender a certain priority by putting something in place until something better comes along. And not only does a presumption of positivity prevent our being altogether empty-handed but it assures that what we get is the most favorable resolution that is rationally available in the circumstances.

4. An Historical Illustration: Cartesian Epistemology

Let us return to Descartes for a moment. His *Discourse on Method* provides us with a striking illustration of the contrast between methodological negativism and positivism. With regard to epistemic matters of knowledge and belief, Descartes took a drastically negativistic line:

> But as regards all the opinions which up to this time I had embraced, I thought I could not do better than endeavour once for all to sweep them completely away, so that they might later on be replaced, either by others which were better, or by the same, when I had made them conform to the uniformity of a rational scheme. And I firmly believed that by this means I should succeed in directing my life much better than if I had only built on old foundations, and relied on principles of which I allowed myself to be in youth persuaded without having inquired into their truth... This made me feel that some other Method must be found... so, instead of the great number of precepts of which Logic is composed, I believed that I should find the four which I shall state quite sufficient... The first of these was to accept nothing as true which I did not clearly recognise to be so: that is to say, carefully to avoid precipitation and prejudice in judgments, and *to accept in them nothing more than what was presented to my own mind so clearly and distinctly that I could have no occasion to doubt it.* (*Discourse on Metaphysics*, Pt. II; tr. Haldane & Ross.)

On this basis, Descartes was prepared to take no cognitive risks. Intellectually he is from Missouri – he will take nothing on trust and requires a personally convincing ("show me") demonstration of anything he is to accept.

Interestingly enough, however when it comes to matters of action and the principles of practice, Descartes takes on emphatically more positive line:

> As it is not sufficient, before commencing to rebuild the house which we inhabit, to pull it down... unless we have also provided ourselves with some other house where we can be comfortably lodged during the time of rebuilding, so in order that I should not remain irresolute in my actions while reason obliged me to be so in my judgments, and that I might not omit to carry on my life as happily

as I could, I formed for myself a code of morals for the time being which did not consist of more than three or four maxims, which maxims I should like to enumerate to you. The first was *to obey the laws and customs of my country*, adhering constantly to the religion in which by God's grace I had been instructed since my childhood, and *in all other things directing my conduct by opinions the most moderate in nature, and the farthest removed from excess in all those which are commonly received* and acted on by the most judicious of those with whom I might come in contact... And...it seemed to me that it was most expedient to bring my conduct into harmony with the ideas of those with whom I should have to live. And amongst many opinions all equally received, I chose only the most moderate, both because these are always most suited for putting into practice, and probably the best (for all excess has a tendency to be bad), and also because I should have in a less degree turned aside from the right path, supposing that I was wrong, than if, having chosen an extreme course, I found that I had chosen amiss. (*Ibid*, Pt. III.)

When philosophy students first study Descartes classic tract, they are almost invariably induced by their instructors to regard this sort of epistemic negativism as something natural, sensible, and desirable and to deplore the practicalistic positivism to which Descartes inclines in some discussions of epistemic fundamentals as something of a misfortune, a grudging and self-protective concession to the obsolete demands of the traditionalistic mentality of his environing culture. It certainly is not generally brought home to them that Descartes' epistemic negativism plunges him into so deep a pit of cognitive vacuity that he requires the aid of a benevolent deity to get him out. Nor is it duly stressed that consistency is a virtue and that and of positive-minded approach is to be adopted with respect to the principles of action it makes sense to adopt one also with respect to the principles of belief – that what is sauce for the goose of *practical* action is also sauce for the goose of *cognitive* action (that is, belief formation) in specific. Descartes never explains – and is not really able to explain – why the evident appropriateness of holding on to our practices unless and until such a time that a discernibly better practice is at hand should not hold with respect to the putative correctness of belief every bit as much as for matters of the putative appropriateness of action.

The methodological consideration that Descartes so sensibly adopts with respect to the appropriateness of practices can and should hold every bit as much where the matter of the appropriateness of beliefs are concerned. "Hold to what you have until something discernibly better comes along" makes perfectly good sense on both sides. In either case there is no good reason to abandon what we have unless and until we have a better alternative that is not only thinkable or imaginable but actually becomes available in some realistic and proclaimable way. Descartes sceptical negativism with regard to cognitive matters is not something self-evidently natural and sensible – as subsequent generations of philosophy professors seem to think – but would make sense only if it has ultimate epistemic fruits to an extent that is decisively counter-indicated by Descartes' ultimate recourse to a deus ex machina to save the epistemic day.

5. Wider Implications

It is only too clear that, in epistemology, positivity militates decisively against radical scepticism. The sceptic is prepared to deny anything we affirm. And when we adduce supporting considerations he simply repeats his initial challenge: "But what justifies you in maintaining *that*?" But of course a presumption of positivity pulls the rug out from under this strategy. If we are in a position to offer something on the basis of an observation that enables it to stand unless and until something else comes along to displace it, then all-out scepticism will not do. For to displace that presumption, the sceptic would – contrary to his approach – have to advance some substantive correlation to serve a purchase-hold for his objection. His tactic of throwing doubt upon everything-all-at-once becomes unavailable.

It is not difficult to see that the presumption of positivity cuts a wide swath across the philosophical domain.

Ethics. In matters of ethics and moral theory, a nihilistic relativism has come to exert an increasingly pervasive and destructive influence. It is based on the idea of an indifferentist egalitarianism seen as pluralism of ethical principles and standards – the variety of positions that we encounter here – as having a mutually annihilative effect. Where we consider some group that holds a position different from ours, then – so the relativist argues – we must consider what they think as being on all fours with what we think. Ergo we have no basis from which to criticize them – and thus no rational basis for maintaining our own position. But a presumption of positivity plays havoc with such an indifferentistically egalitarian approach. If the principles and standards that we have remain place until displaced because a cogent case comes to be made for some alternative, then we always do have a basis for our own judgments. The mere existence of alternatives as such exerts no destructive force. It is not enough that those other positions merely *rival* ours, they must actually *overpower* them.

Axiology. An analogous situation holds with respect to matters of evaluation in general. Here too relativistic negativism runs afoul of the presumption of positivity that operates in favor of the standards we actually have in place. To be sure, it is unquestionably the case that we can only make meaningful evaluations on the basis of standards and criteria that we accept. But we need not beg the question by insisting that our standards are the definitely correct ones. All we need suppose is that owing to a presumption of positivity those standards stand in place unless and until something discernibly better comes along to displace them.

Metaphysics. Metaphysical discussions have long witnessed a prominent minimalism geared to the use, in and out of season, of Occam's razor: *Entia non multiplicanda sunt praeter necessitatem*, or (freely) "Do not countenance entities except when absolutely necessary." Under the aegis of this principle, materialistic nominalists like Armstrong and phenomenological empiricists like Carnap have alike insisted on the sufficiency of their own favored ontological principle. Their line is: "If you can possibly get by in your ontology with one single kind of thing – however complex and convoluted the explanatory mechanisms by which you maintain its sufficiency – then this mechanism is validated by Occam's principle. But is it evident that an ontological approach predicated on the presumption of positivity will take a very different line. "Do not eliminate entities capable of doing useful explanatory service unless and until such a point is reached where other alternatives are available that can do the job equally efficiently." And "efficiency" must now be understood in an all-in-all sense that encompasses not only *ontological* parsimoniousness but also *explanatory* simplicity. A principle of economy is still operative. But this is not just an *ontological* economy of entity types but one that also looks to the *cognitive* economy of explanatory mechanisms. We take once again the optimistic line that the most serviceable of available alternatives deserves our favor.

In sum, the presumption of positivity is capable of doing yeoman work in a wide variety of philosophical contexts.

6. The Question of Validation: The Methodological Basis of the Presumption of Positivity

But must not such a presumption itself be based upon supportive evidence? Must it not be established evidentially that, other things equal (or other things vacuous) the presumptive alternative is the most probable or promising one?

Not necessarily – or at any rate not in the first instance. Presumptions are by their very nature the sort of cognitive resource that need not be based on evidentiation but can be justified on the grounds of their utility. And this is true, in particular, of a presumption of positivity, precisely because of the unproductiveness and sterility that attends to a less than optimistic approach in this regard. After all, a presumption is something *methodological*, not *substantive*. And a methodological principle can be validated on methodological grounds. And here *utility* is a perfectly cogent (appropriate) guide. Our methodological resources *can* be based on evidence, but they need not be. Utility – effectively serving the purposive teleology of the domain at issue can be a perfectly appropriate ground for holding on to something until something better comes along. Both *efficacy* (evidentiated ex post facto an application) and *utility* (manifested in methodological ventures like implicitly) are valid considerations on validity presumptions.

Consider an analogy. You meet a stranger with whom you wish to communicate. You speak four languages – in order of ability English, German, Spanish, and French. You have no idea if he speaks any of them – no assurance that you will be able to communicate with him at all. But in making the trial at starting a conversation, you might as well start to implement your own convenience by working down your competency list. That way you have the assurance of optimization – of assuming that if you do achieve your objective of communication at all, you will do so on the most favorable terms.

Matters stand analogously with the methodological presumption of positivity. In rational inquiry – in the endeavor to obtain sensible and substantial prior guarantees. But if we pursue our question-resolving efforts on the basis of a presumption of positivity, we do have the a priori assurance that if one can obtain a viable answer at all then we shall do so on the most favorable terms – that we will resolve those questions in the most satisfactory way. The justifying rationale of the process is thus that it provides a route to the optimum, the best realizable result – that it represents the strategy which will secure for us the best resolution available in the circumstances.

With methodological optimism we have a decisive reversal of cognitive perspective – a profoundly different methodological attitude from traditional negativism. Its variant standpoint runs something like this: In our endeavors at problem solving, let us stick with what we have until something discernibly better comes along. Why worry about problems until they arise? Sufficient unto the day are the evils thereof – so let's not cross the bridge of further problems and difficulties until we actually get there. Methodological optimism represents a positive approach whose very positivity – and economy of effort – is of itself a powerful asset to the enterprise of rational inquiry and deliberation.

What validates the problem-solving approach at issue in methodological optimism? What justifies our taking this particular procedural line?

The pragmatic approach to validation is naturally attuned to and appropriate for methods and methodological devices. After all, a method is always a method-for: methods, after all, are means for realizing certain sorts of objectives and bear the aspect of purposive efficacy. Accordingly, the issue of efficacy and effectiveness always comes upon the scene. To justify the use of a method we must either proceed categorically (by this-or-nothing argumentation) or comparatively (by competitive advantage argumentation). (The former argumentation can proceed via general principles, the latter requires substantive/factual inputs to establish operative superiority.) But the critical point overall is that the natural and appropriate answer to the rational validation of anything in the methodological realm is though "pragmatic" considerations that turn on issue of functional suitably (efficacy and effectiveness).

After all, in any situation of problem solving the sensible thing is clearly to begin by testing out the most promising-looking possibilities and then proceeding successively to those that are less so. And there is simply no good and sufficient reason for not proceeding in just this same way in the case of cognitive problem solving by granting primacy and priority to the less sure. But of course the instrumentalities by which we achieve this prioritizing in the cognitive case are exactly the factors of plausibility, presumption, and burden of proof. In the cognitive case these "most promising-looking possibilities" are thus exactly the ones to which we should accord the benefit of favorable presumption. With such positivity as with other sorts of presumption we do not assume (universality) that the beneficial is tantamount to substantive correctness. The validation of this presumption is not predictive/evidential but rather is a matter of methodological utility.

7. The Factual Aspect

But what of the factual presuppositions that are involved? After all, methodologies invariably involve some presuppositions about how things work in the world. (Be they methods for transport or for problem solving or whatever, no methods can work effectively *in vacuo*, without the realization of some presuppositional conditions regarding relevant facts.) It is accordingly necessary to come to terms with the fact that our methods and methodological positions are seldom, if ever, *purely* methodological. They almost invariably have substantive components (presuppositions, implications) by way of factual commitments.

Now in this regard the important and noteworthy fact is that *the rational justification of adopting a method carries in its wake a subsidiary (indirect, oblique) justification of its factual components*. And consequently these pragmatic considerations that justify a method's adoption thereby constitute a good reason (albeit of a *practical* rather than necessarily *evidential* reason) for endorsing its factual components.

To be sure in the dialectical interplay of reasons pro and con those practical reasons for endorsing the factual thesis at issue can become overshadowed by evidential reasons to the contrary. If there is definite factual evidence *against* those factual components, then this, of course, tells against them and against the method to which they attach. But in a situation where there were no such evidential obstacles, the practical considerations at issue can do their work unrestricted, and can accordingly support (albeit not thereby adequately *evidentiate*) the factual components of the methods in question. And in particular, as long as the status of those factual components remains indeterminate (unknown) as far as the actual evidence that we can bring to bear on the question is concerned, we reach a situation where purely practical considerations can so operate as to indicate the provisional acceptability of those factual claims. Such supportive considerations operate at any rate pro tem, prevailing unless and until overcome by other considerations to the contrary (what way, of course, be evidential).

Accordingly, the pragmatic considerations that validate our acceptance of a particular method (be they categorical or comparative) will cast a substantive shadow. Insofar as they validate (albeit obliquely and pro tem) the acceptance of the factual component of those methods they carry in their wake – at least tentatively and provisionally – and provide the basis for a rational commitment to these factual theses.

Take an example from the realm of cognitive methods. It is plausible to take the line that the scientific method calls for presupposing the "uniformity of nature" – that nature's *modus operandi* today will reason in place substantially unchanged tomorrow. But the *methodological* validation of scientific method simply carries this presupposition on its back.

Again, take an example from the case of methods of moral reflection. Leading a satisfying life requires a rejection of paranoia, a preparedness to proceed on the principle of trust – that people-in-general are (for the most part) honest, reliable, etc. and that the people we encounter in life therefore merit a presentation of trustworthiness – i.e., that we can trust them as such unless and until there is evidence to the contrary. But this is not something we need to *pre*-establish. The initially methodological validation at issue allows us eventually to re-validate this presupposition retrospectively – with the wisdom of hindsight.

Considerations such as these indicate how methodological considerations can speak for factual claims. They allow us to proceed initially at the level of pre-justifications casting a hopeful galance to the eventual prospect of retro-justifications.

These deliberations have important implications for methodological optimism.

Methodological optimism clearly has factual ramifications. That the world's arrangements suit and favor our quest for ultimately adequate knowledge is a factual presupposition of methodological optimism's *modus operandi*. And this is all too plainly something about which, from the evidential point of view, the returns are not yet in. But we are, nevertheless, rationally entitled to endorse this standpoint (provisionally and pro tem) on the basis of just exactly those methodological considerations that have just been canvassed. For the pragmatic advantage of a methodologically optimistic approach speaks for accepting those factual ramifications that are involved – at any rate pro tem. (In the long run, to be sure, the facts must speak for themselves.)

Rational justification is a complex and many-sided business. Different approaches suit different domains. To justify substantive beliefs we need *evidential* considerations to justify acts we need *teleological* (ends-oriented) considerations, and to justify methods we need *functional* considerations, that is, considerations of this efficacy or suitability. And here, in the absence of impeding counter indications, even considerations of convenience can serve to good effect – at least in the first instance, before the course of experience has developed to the point where other (evidential) sorts of justificatory considerations can be brought to bear.

And these considerations, which hold for methods is general, will also hold for methods in the cognitive domain. We can deploy with respect to our *cognitive* methods the same testing procedures that hold for the validation of methods in general. And it is clear, accordingly, that our resources in the scheme of cognitive methodology can, like all methods, be validated on the basis of their utility in anticipation of an eventual recourse to considerations of efficacy. On this basis, it emerges that methodological positivism is indeed a useful and appropriate tactic. For one thing, the dogmatically negativistic approach "allow no presumptions – accept nothing until and unless it has been established evidentially" is a frustratingly sterile policy that does not permit the process of cognitive problem solving to get off the ground. More serious yet is the impetus of the consideration that if we can get a satisfactory answers at all in the cognitive enterprise of rational inquiry, then an optimistic approach will get us there more straightforwardly – more effectively and at less cost of time and effort – than other, more oblique alternatives. Negativism is a demonstratively barren policy, and it is to all intents and purposes an a priori fact that in methodological positions has the advantage of feasibility and sensibility in the context of rational problem solving.

In sum, what speaks for methodological optimism is exactly its positivity – the fact that insofar as good results are to be had, this way of proceeding – on all available indications – the most effective way to get them.

NOTES

1. Richard Whately, *Elements of Rhetoric* (London and Oxford, 1929), Pt. I, Ch. III, sect. 2.
2. On the strictly empirical side, it is difficult to exaggerate the extent to which our processes of thought, communication, and argumentation in everyday life are subject to established presumptions and accepted plausibilities. This is an area that is only beginning to be subjected to sociological exploration. (See, for example, Harold Garfinkel, *Studies in Ethnomethodology* (Englewood Cliffs, 1967). A pioneer work in the field was that of Alfred Schutz, *Der sinnhafte Aufbau der sozialen Welt* (Wien, 1932).)

CHAPTER 7

MEANINGLESS NUMBERS

> "Deus fecit omnia in pondere, in numero, et mensura."
>
> – Old Testament

> "The business of pinning numbers on things – which is what we mean by measurement – has become a pandemic activity in modern science and human affairs."
>
> – S.S. Stevens,
> "Measurement, Psychophysics, and Utility"

1. THE PROBLEM

Notwithstanding an extensive literature on the subject, the existing state of the art is such that our understanding of exactly what measurement is all about still leaves much to be desired.[1] The present discussion will focus on the problem of distinguishing genuine measurements – quantity-specifications that are informatively significant and meaningful – from those sorts of quantifications that lack this sort of substance. Its problem domain is that of the question: What sorts of quantifications are not just feckless number-juggling but actually *measure* something?

Many treatises on the subject routinely assume that each and every observationally based quantification represents a measurement.[2] But this is nonsense. Quantification as such does not automatically constitute measurement. To see this, consider some examples of effectively meaningless quantities:

- the number of 3's in the distance (in kilometers) between two cities;
- the number of times (on average) that the sentences of a given English text end with a proper name.

It is very doubtful that such numbers do any measuring. For surely not every quantity represents a measurement. "How many of those girls remind you of your mother?" you ask me. "Two," I respond. A lovely quantity, that! But what in heaven's name am I *measuring*? It is thus all too clear that we must reject the claim of S.S. Stevens that "measurement [is] the assignment of numerals to objects or events according to a rule – any rule."[3]

Philosophers of science all too often sidestep the problem. They incline to ease their task by simply supposing that the problem of distinguishing between meaningless quantities and actual measurements will go away if we simply ignore it. But this tactic, however convenient, is very questionable. A measurement, after all, has to be a quantitative characterization of some meaningfully descriptive facet of reality, as opposed to one that is arbitrary and uninformative.

But exactly what is at issue here? A well-known pair of philosophical authors has told us that measurement consists in "indicating the quantitative relations between [intensive] qualities."[4] But this idea that quantitative measurements must represent actually qualitative features of things is deeply problematic. Presumably the rate of exchange between dollars and yen constitutes some sort of measurement. But it is far from clear that the qualities or attributes of something are at issue. Again, birth rates or inflation rates do not discernibly reflect qualities of anything – unless we blatantly *invent* something (a social or economic or political *system*) for them to be qualities of.[5] We can measure the annual snowfall of a place, but it is far from evident that one of its *qualities* is at stake.

Perhaps, then, one should abandon any reference to qualities in this context. Perhaps getting numbers is all that counts for measurement. Following in the footsteps of the operationalist school of P.W. Bridgman, the British philosopher of science Herbert Dingle has insisted that "instead of supposing a pre-existing 'property' which our operation measures, we should begin with our operation and its result, and then if we wish to speak of a property (which I do not think that we should do) define it in terms of that."[6] But here we are caught in a dilemma. If we tie measurement to specifically physical processes of quantity determination – linking it to apparatus manipulation and instrument pointer readings as with the measurement of length or mass – then we proceed in so restrictive a way that we have difficulties accommodating the sorts of quantities at issue with social affairs. Interest rates or the velocity of money circulation are some examples. In macroeconomics, after all, we get our quantities not by reading off the position of a pointer from a scale, but by copying suitably related numbers from pieces of paper. And yet the claims of these quantities to count as measurements seems to be conceded on all sides. It is thus very problematic to insist that measurements have to result from a *physical* measuring process of some sort. If, on the other hand, if we loosen up the linkage of *measuring* to *measurement* too much, we lose our cognitive hold on what measurement is. In particular, if we reject the distinction between genuine *measurements* and merely *meaningless* quantities altogether – if the phenomenology we take into view is that of indiscriminate number assignments – then there at once ceases to be any real reason why anyone should be interested in the topic of measurement at this absurdly general and indiscriminating level.

But how are we to understand this difference between real measurements and meaningless quantities? It would be all very well to say "I know a measurement when I see one," but this convenient approach facilitates understanding no more in the present context than in any other. In the absence of a sensible answer to the question of what it is to measure, there is clearly a large hole in our understanding of the scientific enterprise.

2. Quantification Versus Measurement: What Makes a Number Meaningful?

One sensible idea, it would seem, is that the only sort of quantity specification that deserves the name of measurement is that which plays an *ampliative* role in enabling us to extend our information regarding the items whose quantitative aspects are at issue. Thus contrast the preceding examples of problematic number allocation with such items as:

- inches of rainfall per annum (at a particular place);
- the number of inhabitants (of a particular town).

The situation here is very different. The salient difference seems to be one of cognitive utility. Information of this sort is genuinely *informative*. Given our knowledge of how things work in the world, we can draw various informative conclusions from it – regarding agricultural potential in the case of rainfall (say), and requirements for food and water in the case of population.

On this basis, the meaningfulness of quantities is clearly a reflection of the extent to which they are bearers of information. And it is evident that different albeit interrelated considerations are at work in making a number meaningful in this regard. An adequate inventory is bound to include the following:

2.1. *Effective Determinability*

An actual measurement has to be the result of a practicable and implementable process. And this is simply lacking in various cases. Consider, for example:

- the number of Latin words whose meaning a person once knew but has forgotten.

Or again, for living individuals contrast years from birth with years to death – that is, with

- a person's *remaining* lifespan.

The problem with such quantities is that they are effectively impossible to determine. We can make sense of the relative frequency with which a certain word is used by the writers of a certain language or the number of times their written discussions refer to a certain person, but there is little if anything we can do with

- the percentage of persons to whom a certain idea has occurred.

We cannot plausibly construe such problem quantities as meaningful measurements. Clearly, measurement has to be an operationally implementable procedure of some sort.

2.2. Reproducibility

An actual measurement has to be well defined – the stable result of a reproducible process and procedure. Real measurements must yield essentially the same result on different occasions when carried out by the same operator, and must also yield essentially same results when carried out concurrently by different operators. This requirement is illustrated by contrasting the number of "wrongdoers" with "the number of people found guilty of specific offenses (from some specified list)." Or again, contrast the number of unhappy people in a community with the number of suicides or the extent of its purchase of headache remedies.

2.3. Context Invariance/Robustness

A genuine *measurement* should reflect a substantially context-invariant quantity. Since talent and skill for taste performance is generally comparative and context variable, industrial statistics regarding "skilled" versus "unskilled" workers are automatically suspect – particularly because the same product can be made from the same materials by processes and methods requiring very different sorts of talents.

In particular a genuine measurement must not be sensitive to factors which – as best we can tell – are not causally linked to the sort of thing at issue. For example, it just does not – and should not – matter whether we measure the amount of rainfall at a certain place on an odd or an even-numbered day, a workday or a holiday. In quantifying highly context-dependent relational and interactional aspects of things – say hair vibration as opposed to pulse rate – we are not in general carrying out actual *measurements*. (Hair vibration will vary with wind exposure, body movements, headscratching, etc.)

2.4. Validity/Coherence

With genuine measurements, there must be good reason to think that the result of the measurement is the true value of the quantity at issue. And this is only possible where measurements cohere. Wherever there are different measuring processes and procedures – different ways of measuring what is supposed to be the same item – they must agree. Thus whenever the operators and their mode of operation themselves enter into the measurement process in a result-influencing way (as some physicists claim for the "measurements" of quantum phenomena, but as certainly holds for various sociological quantifications), the claims of the resulting numbers to represent measurements are *ipso facto* compromised.

2.5. Predictive Utility

It is clear that the usefulness of a number in predictive contexts is a pivotal feature of its meaningfulness. If we know a town's population we can use this datum to say something about this locality's consumption of water or its need for housing, while the age of its mayor or the average velocity of its winds from the east are, by comparison at least, meaningless quantities because few if any other quantities relevant to the constitution of towns can be predicted on the basis of these data.

This factor of the predictive serviceability of measurement is closely linked to yet another crucial aspect of measurement, namely:

2.6. Nomic Involvement

Measurements should result in quantities that function informatively in the context of general laws or law-like statistical relationships. This means, in particular, that, while any mathematical compounding of quantities will of course yield yet another quantity, there is no cogent reason for thinking that this is so with measurements. Even when the input quantities actually measure something, the functionally combined output quantities need not do so as well, owing to an absence of informative interrelationships. We would thus be highly disinclined to see

$$\sqrt{\text{age (in years)} + \text{weight (in kg)}}$$

as *measuring* some aspect of a person's make-up. Given this quantity, there is effectively nothing else of any interest with which it stands in lawful interrelationships.

It is tempting to distinguish between "fundamental" measurements which we carry out "directly" by some suitably contrived process of physical manipulation with material objects and "derived" measurements which we effect by *calculations* with numbers resulting from prior measurements.[7] But – as the previous examples show – even we put perfectly good measurements to work, it is clear that their arbitrary arithmetical compounding does not automatically measure anything. And this circumstance blocks the prospect of any *recursive* approach to the specification of legitimate measurements.

In this light, it is the *laws of nature* that ultimately determine the meaningfulness of complex quantities. Consider the gas law to the effect that (at least approximately) the product of pressure and volume divided by temperature is constant:

$$(P \times V)/T = \text{constant}$$

Given this law, the product $P \times V \approx T$ represents a perfectly meaningful quantity that provides us with a means of measuring temperature. But the quantity

$$P \times T = PT^2/T \approx T^2/V$$

fails to measure anything because it lacks any plausible physical meaning owing to its failure to operate as a significant whole in the context of serviceable laws. As this example shows, when we combine perfectly meaningful and measurable quantities by throwing them together arbitrarily via rules of calculation, the upshot may well be something altogether nonsensical. The result of a real measurement has to be something that functions as a meaningful unit in the context of laws.

Genuine measurements accordingly have to be theory-laden: when we cannot embed quantities in a theoretical context of some sort and fit their results into our theories, then these number specifications are clearly unable to qualify as authentic measurements.[8] And if the theories into which they fit are patently invalid, absurd, or crazy, then the quantities at issue cannot qualify as genuine measurements. The range of evil-eye potency or the intensity of a person's "spiritual aura" are putative measures that vanish into nothingness with the disappearance of the theories on which they are predicated.

On the other hand, the mere fact that quantities do indeed behave in lawful coordination is of itself not enough to mean that they represent meaningful measurements. Thus consider once again the gas law:

$$P \times V \approx T$$

And let

$$P^* = P \times M$$

and

$$T^* = T \times M$$

where M represents some otherwise irrelevant feature of the system at issue – its aggregate mass – or, if you prefer, the time of sunrise in its location. Then those star quantities are clearly rubbish. Yet nevertheless, they stand in splendid lawful coordination via the gas law. As this example indicates, the law-connectedness of determinable quantities does not suffice to establish them as meaningful measurements.

We arrive at the recognition that each of the six measurement-characterizing factors considered individually – determinability, reproducibility, robustness, nomicity, and the rest – has its problems. More ominously yet, it emerges that, even in combination, they do not suffice to assure meaningfulness. They are all *necessary* for genuine measurement, but even jointly they are still not *sufficient*. Consider, for example, the following person-descriptive parameter:

- the sum of a person's age in years plus their systolic blood pressure.

Not only is this well-defined, determinable, and contextually stable, but it also has nomic involvement and predictive value in relation to such issues as general health and life expectancy. All the same, it is not something we could regard as a meaningful measurement because of the rather eccentric way in which it mixes apples and oranges. As such considerations indicate, something more is needed – over and above all of the preceding factors – to establish the claims of a quality to representing a meaningful measurement. A clearly appropriate further step moves in the direction of:

2.7. Dimensionality

Actually to *measure* something is to affix a numerical yardstick to some quantitative *parameter* that has its operative foothold in the world's scheme of things – length, temperature, mass, electric charge, money circulation or the like. Measurement must present an objective and well-defined quantitative aspect of the qualitative make-up of things in the real world. A measurement, after all, is a number assignment made under a particular description such as:

- is x inches long;
- measures x units on the Richter scale;
- rates x units of acquisitiveness on the Spencer phrenology scale.

And when numerical assignments fail to capture such lawfully descriptive features of things, then they just do not *measure* anything.

With a genuine measurement two questions arise:

(1) *What* is it that one is measuring? (The *object* question)
(2) *How* is it that one is measuring what is at issue? (The *process* question)

And, moreover, the issues involved in (1) and (2) must be separable and distinct – that is, one must in principle be able to provide an answer to the *what* question independently of one's answer to the *how* question. Thus a person's weight or age represent cogent measurements, but not:

- age in years divided by waistline in inches;
- height in feet plus years resided at present address.

It is clear in this context that a wide roadway leading from meaningful to meaningless numbers is provided by mathematical compounding. This is indicated by such examples as:

- the product of the longest and shortest side of a polygon;
- weight of a person in kg minus months resided at present address;
- volume of an object (in cc^3) divided by its age (in years).

Such amalgamations are problematic precisely because there is no noncircular reply to the question: Just what is it that is being purportedly measured in this way? That is, we have no workable way of distinguishing the *what* of the putative measurement from its *how*.

The issue of dimensionality is clearly serviceable in providing a cogent factor disqualifies such Rube Goldberg quantities from counting as genuine measurements. Regrettably, however, it is not easy to say just what is at issue here (which may explain why recent treatments of measurement in general simply bypass this issue of descriptive dimensionality). Philosophers of science have in recent years deliberated a good deal about what is or is not a *natural kind* when it comes to *classification*, but they have largely ignored the closely parallel and inherently no less important issue of what constitutes a *natural dimension* in point of *mensuration*.[9] Yet however difficult this issue may be to resolve, it is clear that such descriptive dimensionality – or something very like it – must be added to our list of requisites.

Our deliberations have indicated that the following features all represent necessary conditions for meaningful measurements:

- effective determinability;

- reproducibility;
- context stability/robustness;
- coherence/validity;
- nomic involvement;
- predictive utility;
- descriptive dimensionality.

The yet (unresolved) problem we face is whether now at last these several necessary conditions for genuine measurement are jointly sufficient. One may be tempted to surmise that this is indeed the case. But there is good reason to think that it is not. For one thing even perfectly meaningful quantities can be contextually problematic. It makes sense to ask for the market price of gold or of lead specifically but not of metal in general. We can make good sense of the idea of the average birth weight of a human female in specific, but can make little of the color of the average death weight of a fish in general. But an even graver difficulty lurks around the corner. Consider, for the sake of an example, the person-correlative quantity:

$$\text{fage} = \text{age} \times \frac{\text{\# of fingers}}{\text{\# of toes}}$$

This clearly quantifies an aspect of a person's individual make-up in a way that (1) has a meaningful dimension (viz., chronological age, seeing that the fractional multiplier is dimensionless), and (2) behaves lawfully since from the vast majority of humans *fage* = *age* and a person's age factors lawfully in many contexts and as biological development and life expectancy. Despite all this, however, the Rube Goldberg nature of the conception would leave one disinclined to consider "fage"-determination as a meaningful measurement process. What we want and need is a more cogent and discriminating account. But it is one thing to ask for such an account and another to know where it can be found.

And so, in the end, it appears that we are confronted with a distinctly discomfiting situation. It is clear *that* we need a cogent way of distinguishing between meaningless quantities and genuine measurements. But it is far from clear *how* we are to draw this distinction.

In the philosophy of science, there are a number of structurally analogous problems of distinction and demarcation – between natural kinds and haphazard collections, between genuine laws and accidental generalizations, between ad hoc explanations and genuine ones, and even between real science and pseudo-science. The problem of distinguishing between mere quantification and real measurement is of this same general sort. But unlike the rest, it is a problem that has received precious little attention in the literature of the field. In this case even more than with others we have to do with an instrumentality of communication about science that is clearly useful, but which, nevertheless, we are simply unable to render clear and precise in anything like a satisfactory way. On all indications, the distinction between actual measurements and meaningless qualities remains enigmatic.

Some numbers can be acknowledged as measurements because, like weight and distance, they are paradigmatic of the very concept. Others are clearly not measurements because they violate one or another of the necessary conditions of the conception. But there is a considerable grey area where we do not see the way clear, and where we have good reason for caution and unease. And nowhere does this lack of a clear-cut way of drawing the line manifest itself more painfully manifest than in the social sciences. Let us explore some of the ramifications of this situation.

3. PROBLEM MEASUREMENTS 1: INTELLIGENCE

The need for maintaining the distinction between *measurement* and mere *quantification* along the lines of the preceding deliberations, this raises some disturbing questions regarding the status of various putative "measurements" that people nowadays conjure with in the social sciences – particularly in areas that have implications for and applications to matters of practical policy. In economics, for example, we deal with quantities – such as interest rates – that are neither determined by actually measuring anything, nor yet by calculation from such quantities, but which emerge from calculations with numbers that crop up – mysteriously or otherwise – as writing on bits of paper. The idea of a meaningfully descriptive dimensionality in this domain is particularly problematic.

And calculating with such quantities poses further problems. As our preceding deliberations have shown all too clearly, when one combines different quantitative factors by mathematical compounding one will generally get not a measurement but a mess. And while this mess may possibly behave lawfully in some respects, nevertheless from the angle of *understanding* – of facilitating a conceptualized grasp of things – it remains a mess unless and until it comes to play a substantial role in a systematic law framework of theoretical systematization.

This state of affairs casts a dark shadow over some of the quantities used in economics, but particularly over many of the quantities used in social analyses and social studies, where people throw numbers together in ways that ought to raise one's theoretician's hackles. A good illustration of such a problem number is provided by the IQ tester's idea of a pervasive "intelligence." How cognitively competent people are is clearly of interest to us in many contexts. For, all too evidently, people have very different levels of ability – be it bodily or mental. But there is no good reason to expect that someone's physical or intellectual dexterity or versatility can be represented by a single number. And the fact that it would be *convenient* if this was so from the angle of our programs of educational planning and management just does not make it true. Intelligence quotients throw together ability to decode language with analogy spotting and competence at calculation. Those tests explore very different sorts of abilities, and there is no reason to think that these can simply be aggregated by weighted addition to yield a meaningful composite result. Even a student's performance on one of those objective tests, on which we Americans place so much weight in college admissions processing, and in scholarship awards, actually indicates little about ability for intellectually demanding work – or for doing anything substantially different from taking objective tests.

4. Problem Measurements 2: Quality of Life

For another example of problematic qualification in the social domain, consider the nowadays fashionable quality of life measures for different towns, areas, or countries. Here income data, housing costs, crime statistics, cultural facilities are somehow quantified and then thrown together in some sort of aggregation. But there is clearly no good reason to think that a meaningful result emerges. Conceivably such QOL figures could be used to predict migration patterns, although this has apparently never been attempted. ("Other things equal, people abandon low QOL areas for high QOL ones" is hard to test and unlikely to survive testing in unqualified generality.) But clearly some embedding in a structure of lawful order is needed, and has never been supplied. And there is an even more fundamental problem here.

In general, things have many different value aspects and we have no workable single-valued "function of combination" enabling us to extract a single, all-embracing *measure* of overall value from them. We have to deal with a plurality of distinct considerations that are not easily weighed off against one another. Even if we assume (perhaps rashly) that measurability is possible *within* each of these parametric dimensions, there is generally no way of making quantitative exchanges across different value parameters by way of weighing them off against each other in a common scale. The elements of a good journey are clearly not commensurable: adding more spectacular scenery cannot make up for bad food, even as adding more salt, no matter how much, will not compensate for a cake's lack of sugar. We can readily quantify various features of urban life – crime rates, housing costs, park acreage per capita, and so on. But does that mean we can combine such numbers into one single meaningful index of quality-of-life? Surely our prospects in this direction are no better than those of the 19th-century French physicians who thought that "health" was a property of the human body and attempted to provide a measure of a patient's "health" or "constitution" by means of a single number concocted out of a witches' brew of diverse factors.[10] With quality of life, or health or other such factors, the various evaluative aspects of the good are not interchangeable. In general, when we have to appraise a profile or complex of desiderata, preferability becomes a matter of a contextually determined *structural harmonization* rather than of a mensurational maximization.

5. Larger Vistas

In the aggregate, then, these deliberations indicate the need for recognizing that, even when we are dealing with a perfectly fine *quantity*, the conditions needed for this to qualify as an authentic *measurement* may nevertheless not be satisfied. The importance of science in modern life has engendered a quantitative prejudice. People incline to think that if something significant is to be said, then you can say it with numbers and thereby transmute it into a meaningful measurement. They endorse Lord Kelvin's dictum that "When you cannot express it in numbers, your knowledge is of a meager and unsatisfactory kind."[11] But when one looks at the issue more clearly and critically, one finds that there just is no convincing reason to think this is so on any universal and pervasive basis.

Science has succeeded in mathematicizing the realm of our *knowledge* to such an extent that we tend to lose sight of the fact that the realm of our *experience* is not all that congenial to measurement. It is full of colors, odors, and tastes, of likes and dislikes, of apprehensions and expectations, etc., that are not particularly amenable to measurement. We readily forget how very special a situation actual *measurability* is – even in contexts of seeming precision.

Moreover, reliance on numbers brings in its wake a host of problems of its own. For one dangerous thing about numbers is that small errors in their use can produce large – and very unfortunate – consequences. A minor mistake in the number encoding of a prescription medication can prove lethal. Again, for many years, spinach has enjoyed a great prestige as a valuable source of iron because a misplaced decimal point credited this vegetable with an iron content ten times its actual value.[12]

A fetish for quantification seems to be astir among our contemporaries. We worship at the altar of statistics: the penchant for quantities is a salient characteristic of contemporary Western culture. Everything we touch turns to numbers: intelligence quotients, quality of life indices, feminine beauty (ranked on a scale from zero up to a "perfect 10"), and so on. It is thus easy to see why the prospect of meaningless quantities should cause unease. For in this measurement-enchanted time of ours we constantly invoke on quantitative information as a basis for decision making and policy guidance. But garbage in, garbage out. If those quantities that people throw about so readily are in fact meaningless, then the decisions we so enthusiastically base upon them are built on sand. In many situations we would do well to settle for rough-and-ready judgmental appraisals because meaningful quantitative measurement is simply infeasible.

After all, numbers do not always tell the story. The wise strategist realizes that God is not always on the side of the big battalions. "How many divisions has the Pope?" Stalin asked. It was not a question that bemused General Jaruzelski in Poland. By all quantitative yardsticks – body counts and all that – the U.S. was ahead throughout the Viet Nam conflict and duly won all the battles; but it lost the war. One does not need to enroll oneself, with Goethe, as an opponent to measurement's entry into the domain of human doings and dealings to feel a deep disquiet regarding the particular ways in which people have sought to introduce measurement into the social sphere. The fact is that in everyday life, professional practice, and public affairs alike, stubborn reliance on numbers can sometimes prove more of an obstacle than an aid to critical and reflective thought.

The immense success of quantitative techniques in the mathematicizing sciences has misled people into thinking that quantification is the only viable road to cogent information. But think – is it really so? Where is it written that numbers alone yields genuine understanding – that judgment based on structural analysis or qualitative harmonization is unhelpful and uninformative, so that where numbers cannot enter, intelligibility flies away? (After all, modern mathematics itself is not all that quantitative, seeing that it is deeply concerned with issues such as those of topology and group theory that deal with structures in a way that *parts* quantitative issues aside.)

It must be stressed, however, that to acknowledge the limits of measurability is not to downgrade the whole process, let alone to propose its abandonment. It is precisely because we are well advised to push the cause of measurement as far as we legitimately can that we need to be mindful of the line between meaningful measurements and meaningless quantifications. That we cannot draw this line better than seems to be the case at present is – or should be – a proper occasion for justified chagrin.[13]

NOTES

1. A collection of particularly fine articles on the subject is Harry Woolf (ed.), *Quantification: A History of the Meaning of Measurement in the Natural and Social Sciences* (Indianapolis, 1961). See also Brian Ellis, *Basic Concepts of Measurement* (Cambridge, 1966).
2. See, for example, the Introduction to David R. Kranz et al., *Foundations of Measurement*, Vol. I (New York and London, 1971).

3. S.S. Stevens, "Measurement, Psychophysics, and Utility," in *Measurement: Definition and Theories*, ed. by C.W. Churchman and P. Ratoosh (New York, 1959), pp. 18–63 (see p. 19).
4. M.R. Cohen and Ernest Nagel, *An Introduction to Logic and Scientific Method* (New York, 1934), p. 294. Compare Norman R. Campbell: "Measurement is the process of assigning numbers to represent qualities," *Foundations of Science* (New York, 1957).
5. Perhaps one of the reasons for being of the fashionable idea of a "system" is to play just exactly this role.
6. Herbert Dingle, "A Theory of Measurement," *British Journal for the Philosophy of Science*, vol. 1 (1950), pp. 5–26.
7. For this distinction see N.R. Campbell, *Physics: The Elements* (Cambridge, 1920) and also *What is Science*? (Cambridge, 1921); reprinted (NY, 1952.)
8. Recent deliberations in the philosophy of science have done much towards showing that the attribution of observable features to the furnishings of nature is always theory laden – part of a process in which our theoretical understanding of nature plays a significant role. But this is all the more so with respect to the attribution of measurable features, seeing that measurement is even more pronouncedly theory-involving than observation.
9. The issues are closely parallel because a measured feature of something can be looked upon as constituting a parametrized quality of it. As my Pittsburgh colleague Nuel Belnap has noted in discussion, all of the seven factors singled out in the previous quantity-oriented discussion have close analogues in the qualitative case in relation to natural kind versus random assemblage distinction. The difficulty of maintaining natural kindhood in the face of *logical* combinations (such as conjunction) exactly parallels the difficulty of maintaining actual measurementhood in the face of *mathematical* combinations (such as multiplication).
10. See Richard H. Shryock, "Quantification in Medical Science," in Harry Woolf (ed.), *Quantification* (Indianapolis, 1961). Note, incidentally, that "life expectancy" will not do the trick here, seeing that a person may well expect a long life on whose course he often or generally "feels miserable."
11. William Thomson, Lord Kelvin, "Electrical Units of Measurement," *Popular Lectures and Addresses*, 3 vols (London, 1889–91), Vol. I, p. 73.
12. See Peter Skrabanek and James McCormick, *Follies and Fallacies in Medicine* (New York, 1990), p. 27.
13. Other aspects of this chapter's theme are treated in Chapter Eight, "Number Idolotry and Fallacies of Quantification" in the author's *Forbidden Knowledge* (Dordrecht, 1987).

CHAPTER 8

CONCEPTUAL IDEALISM REVISITED

1. Introduction

Idealism, broadly speaking, is the doctrine that reality is somehow mind-correlative or mind-coordinated.[1] However, the specifically *conceptual* idealism which concerns us here stands in contrast to an *ontological* doctrine to the effect that mind somehow constitutes or produces the world's materials. Instead, it maintains that any adequate descriptive characterization of physical ("material") reality must involve an implicit reference to mental operations – that some commerce with mental characteristics and operations always occurs in any viable explanatory exposition of "the real world." The central thesis of this position is that the mind is responsible for nature-as-we-understand-it, not, to be sure, by making nature itself, but rather through its role in the mode-and-manner determining categories in whose terms we conceive of it. On its approach, the constitutive role of mind in nature is thus to be thought of neither in ontological nor in causal terms, but hermeneutically by way of concept explication. It is not that mind produces nature, but rather that the way in which we conceptualize of nature involves the analogy of mind – that we conceive of the real in mind-correlative terms of reference.

Against this background, conceptual idealism's central thesis is that the main characterizing properties ascribed to physical things in our standard conceptual scheme are at bottom *relational* properties, with some facet of "the mind" (or of minds-in-general) serving as one term of this relation. Specifically, it holds that the conceptual scheme we standardly use to construe our experience ascribes to "material" objects properties and characteristics that involve some reference to mental operations within the very meaning of the terms at issue. Even the physical properties of things are yet mind-involving in the analytic sense that their explicative conceptual unpacking calls for a reference to minds and their capabilities.

2. Conceptual Mind-Involvingness

Conceptual idealism is predicated upon the important distinction between conceptual mind-involvingness and explicit mind-invokingness, illustrated in the contrast between a *book* and a *dream.* To characterize an object of consideration as a *dream* or a *worry* is explicitly mind-invoking. For dreams and worries exist only where there is dreaming and worrying, which, by their very nature, typify the sorts of things at issue in the thought processes of mind-endowed creatures: where there are dreams or worries, these must be mind-equipped beings to do the dreaming and worrying. A book, by contrast, seems at first sight entirely non-mental: books, after all, unlike dreams or worries, are physical objects. If mind-endowed beings were to vanish from the world, dreams and, worries would vanish with them – but not books! Even if there were no mind-endowed beings, there could certainly be naturally evolved book-like objects, objects *physically indistinguishable from books as we know them.* Nevertheless there could not be *books* in a world where minds had never been in existence. For a book is, by definition, an artifact of a certain purposive (i.e., communicative) sort equipped with pages on which "reading material" is printed. Such purposive artifacts all invoke goal directed processes of a type that can exist only where there are minds. To be a book is to have *writing* in it, and not just *marks*. And writing is inherently the sort of thing produced and employed by mind-endowed beings. In sum, to explain adequately what a *book* is we must thus make reference to writing and thereby in turn, ultimately to minds. The point is not that the book is mentalesque as a physical object, but rather that to explicate what is involved in characterizing that object as "a book" we must eventually refer to minds and their capabilities, seeing that a book is by its very nature something for people to read. A world in which there neither are nor ever have been minds can contain objects physically indistinguishable from our books and nails, but books and nails they could not be, since only artifacts created for a certain sort of intelligence-invoking purpose can correctly be so characterized. And so, while books – unlike dreams – are not mental items, their conceptualization/characterization must nevertheless in the final analysis be cast in mind-involving terms of reference.

This conception of the mind-involving (as contrasted with the mind-invoking) will be central in these deliberations. For the pivotal thesis of conceptual idealism is that we standardly think of reality in mentalesque terms. It rests on two basic theses:

(1) That *our* world, the world as we know it, is – inevitably – the world *as we conceive it to be*, and
(2) That the concepts (thought-instrumentalities) that we standardly use in characterizing and describing the world contain in their make-up, somewhere along the line, a reference to the operations of mind.

Observing that our "standard conception" of the world we live in is that of a multitude of particulars endowed with empirical properties and positioned in space and time and interacting causally, conceptual idealism goes on to maintain that all of the salient conceptions operative here – particularity, spatio-temporality, causality and the possession of empirical properties – are (conceptual idealism contends) mind-involving in exactly the sense explicated above. (Of course, in speaking of mind-involvement or mind-invocation, no reference to any particular mind is at issue. The mental aspect here operative is not private or personal: it is not a question of *whose* mind – of this or that mind rather than another. The dependence at issue is wholly generic and systematic in nature.)

It is axiomatic that reality, insofar as apprehended by intelligence at all, can be apprehended only through the mediate employment of conceptual mechanisms of representation. And, as conceptual idealism sees it, these mechanisms as we have them here and now are inherently mind-referring. This doctrinal stance holds that the conceptual instrumentalities we use for the characterization of "the real world" are committed in their ideational content to an involvement with actions and accomplishments that minds also can realize. Its pivotal contention is that our world model, the world as we depict it in thought and cognition, is a mental artifact that is constructed (in part) in mind-referential terms – that the nature of the world as we conceive of it reflects the nature of mind. For the salient thesis of conceptual idealism is that we must recognize as mind-involving all of the standard elements of our usual coordinating world-picture, namely:

- particularity/individual identity;
- causality and process;
- space-time emplacement.

All of these factors – identification, causal explanation, and spatio-temporal positioning – are mind-involving activities that envision the world's operations in terms of characteristically mental processes.

It is sometimes said that idealism is predicated on a confusion of objects with our knowledge of them and conflates the real with our thought about it. But this charge misses the point. The only reality with which we inquirers can have any cognitive commerce is reality as we conceive it to be – our only informative access to reality is through the mediation of conceptual resources that are not only mind-supplied but also mind-patterned. Conceptual idealism's thesis is not the trivial one that mind makes the *idea* of natures, it is not open to Santayana's complaint against Schopenhauer that "he proclaimed that the world was his idea, but meant only (what is undeniable) that his *idea* of the world was his idea." Rather, what is at issue is that mind-patterned conceptions are built into our idea of nature: that the way we standardly conceive of nature is in some crucial respects patterned on our mind.

This *conceptual* idealism sees mind not as *causal source* of the materials of nature, but as indispensably furnishing some of the *interpretative mechanisms* in whose terms we understand them. It maintains that we come to cognitive grips with nature on our own terms – that is, in terms of concepts whose make-up involves some reference to minds and their operations. Conceptual idealism is predicated on the view that reality as we standardly conceive it – in terms of material objects characterized by empirical properties and intentions with one another in setting of space and time – is in some respect involved unavoidably in making reference to the operations of mind. As *this* sort of idealism has it, the mind understands nature in a manner that in some ways reflects its own operations in some fundamental respects.

To be sure, the mind-relatedness in view is not something voluntaristic over which persons have power and which can be changed around by their thinking ("wishful" or otherwise). To say something (e.g., a color) is mind-involving[2] is not to say that it is illusory or delusory or otherwise "unreal"; it is not to say that it is psychological in any subjective, person-relative sense; and it is certainly not to say that it is arbitrary, choice-related, or changeable by human discretion. The idealist view of reality is not a do-it-yourself position that lets us shape the world in any way we please. Historically, idealists have never made reality a matter of wishful thinking that lets thought shape the world without objective constraints. Constraints have always been recognized: God (in Berkeley), the faculty structure of the mind (in Kant), biological and perhaps ultimately Darwinian considerations (in Bergson). And conceptual idealism correspondingly acknowledges the restrictive role of an objectively given conceptual scheme. Thus the items we characterize as "mind-dependent" can be perfectly interpersonal and objective; they need not be subjective at all – let alone be something over which people have voluntary control. Moreover, it cannot be emphasized sufficiently that such a conceptual idealism does not take the form of spiritualism or panpsychism; it need not and does not maintain that *only minds exist* and that all there is somehow the work of mind in any productive or causative sense. Rather, this idealist position sees our characterizations of the real as proceeding in mind-referring terms. The position rests squarely on the classical idealistic doctrine that mind contributes essentially to the constitution – as well as the constituting – of our knowledge of reality, achieving this by characterizing things *in its own terms*, in using inherently mentalistic conceptions and categories.

The conceptual idealist sees mind as an explicative paradigm for our understanding of the real, rather than as a productive source in the causal order of its genetic explanation. This position does not see mind as the causal source of reality, but as making a formative contribution to the shape – as well as the shaping – of its characterization of the real.

There may well be a realm of mind-independent reality – indeed a commitment to its existence is unquestionably part and parcel of our standard conceptual scheme for thinking about the real. But while this mind-detached reality underlying the world of our experientially grounded knowledge is something to whose existence (whose *that*) we stand committed but about whose nature (whose *what*) we neither do nor can secure any descriptive information. Reality as we know it is cognitively assessable only through the mediation of our concepts; it is an empirical reality because it is accessible to us through experience alone, and this experience is always *interpreted* experience – there just are no mind-untouched experiential building blocks that the mind first receives and then assembles, because mind from the very start touches and pervades everything experiential. The mind-untouched reality underlying the world of our experentially grounded knowledge is something to whose existence (whose *that*) we stand committed but about whose nature (whose *what*) we neither do obtain nor can obtain any knowledge.

3. Particularity as Mind-Involving

Let us begin the survey of conceptual idealism's main theses by considering particularity. To be a particular is to be identifiable as an individual item with a nature and identity of its own: to be distinguishable as one discrete item in contradistraction to others. But even as what makes something a book is that it is readable, so what makes something an object in the real world is that it is experientiable. In its very nature, particularity consists in identifiability, distinguishability, discriminability. But all of these processes (identifying, distinguishing, discriminating) are fundamentally mind involving; each involves attention-directing and is accordingly the sort of thing that mind-endowed beings – and only mind-endowed beings – can do.

The impact of this argument that identification is mind-involving is tempered by the following line of objection:

> Let it be granted (says the objector) that your argument has shown that to say 'X is *identified*' is to make a mind-referential claim. But this does not mean that 'X is *identifiable*' is mind-involving. Your approach slurs the crucial distinction between actuality and possibility. For consider the pairs: described/describable, mentioned/mentionable, indicated/indicatable, and identified/identifiable. If one grants that the first member of such a pair is mind-involving, one does not thereby concede that the second member is. Thus, saying that a certain particular is *identified* may well carry a covert-reference to a mind, but this does not show that its *identifiability* is mind-dependent. Consequently, since generic particularity demands only identifiability, and not actual identification, your argument that actual identification is mind-involving does not show that *identifiability* is, and so does not suffice to establish the conclusion that particularity is.

This objection is to some extent well taken and specifically so insofar as its aim is to make the point – a surely correct one – that *actual* identification is not a necessary requirement for being a particular, since there is no contradiction in saying that there are particulars which are not identified (though obviously one cannot give an example of one). But the impetus of the objection can be deflected by recognizing the fundamental difference between identification on the one hand, and description, indication, and the rest, on the other. For identification is, in the present context, entirely unique and *sui generis* in a way that impedes straightforward application of the analogy of actual and potential on which the objection rests.

It makes perfectly good sense to say of something that is describable but not described or indicatable but not indicated. The actual/potential distinction is indeed operative in these cases. *But this is not so with identification*: We cannot in principle meaningfully say *of something that IT* is identifiable but not identified, because to say this is to commit a literal nonsense. One would be saying explicitly that one doesn't know what one is speaking of. Until it has been identified (however imperfectly) we simply are not dealing with a particular individual thing: we cannot appropriately be held to say anything about "it" – not even that it is identifiable. To say this is not, of course, to deny that we can speak of otherwise unspecified particulars, as in a statement like "One of the trees in this forest has treasure buried beneath it." But cases of this sort pose no difficulty for our position. For if indeed there is treasure under just one tree, then we have, in effect, succeeded in making an identifying reference to it (as "the tree that has treasure beneath it"); but if there are several trees above the treasure (or none) at all, then there just is no "it" about which we can be said to be speaking: our purportedly identifying reference fails to refer, so that our statement becomes – under these circumstances – semantically unviable.

It is useful at this stage to recall once more the important distinction between the *ontological* mind-dependency of mind invokingness and the *conceptual* mind-dependency of mind involvingness. Granted, only identification is mind-dependent in the strong sense of mind-involvingness, and not identifiably. But that of course does not prevent identifiably from being *conceptually* mind involving – as it indeed is, seeing that the issue pivots on the focussing of attention. To say of something that it is related to minds in a certain way (specifically in the way of admitting being identified by them) is obviously to characterize it in conceptually mind-referring terms (even as describing it as *visible* would be to characterize it in conceptually sight-referring terms of reference).

Again, consider the following objection: "It goes without saying that a critical distinction exists between *the thing identified* on the one hand, and on the other *the identification of the thing*. Your argument overlooks this key distinction. Surely, to establish the mind-involvement of the *identification* of the thing does not show the mind-involvement of the identified thing as such." This objection hits wide of the mark. It rests on a wholly inappropriate comparison. For there is a crucial disanalogy between identifying on the one hand and seeing or describing on the other. We can only consider "the thing identified" once some "identification of the thing" has been given: *any* possible entertainment of the former (the thing identified) is contingent upon the latter (the identification of the thing). By contrast, nothing whatever about the describability of something rests upon the description of this thing. The hypothetical removal of its identification removes the thing in view in a way that the hypothetical removal of its description would not. And this critical disanalogy is fatal to the objection. For individuatability is not a somehow incidental feature of a thing but is crucial to its very existence *as an individual thing*. Consequently, to show that mind-invocation enters in *at this specific point* is to establish that the very standing of a thing *as the individual particular it is* rests upon a resource the very concept of which involves essential reference to the workings of minds.

To summarize: only two conceptual routes lead into the realm of the particular, that of actual identification and that of potential identifiability. The former – identification – is conceptually mind-invoking because identification is an attention-directing, and thus an overtly mental process. The latter – identifiability – is implicitly mind-involving both because of the mentalesque nature of *identification* itself, and also (to foreshadow) because of the mind-involving nature of possibility as such.

The mind-dependence of particularity has far-reaching ramifications. For not only does it affect particular things/substances, it also affects particularity in other categories. Specifically, consider the following:

- the attribution of a *particular* physical property to something;
- the attribution of a *particular* spatio-temporal feature (e.g., shape);
- the specification of a *particular* cause or effect;
- the specification of a *particular* process.

Each of these involves a fundamentally mind-invoking transaction, because in each case the particularity at issue is enmeshed in an invocation of mind that is of just the same general sort that is at issue in item identification.

4. Causality as Mind-Involving

Reality and causality are inseparably coordinated. To be physically real is to be a part of the world's causal commerce. But the conception of causality itself is obviously inseparable from that of causal principles or laws. Now, causal laws are more than mere generalizations. They have counterfactual force and characterize not just what does happen when but what would happen if. The element of causal necessity and counterfactual force is inherent in their nature, and they encompass not only the actual but the hypothetically possible as well. But this realm of mere possibility – of what is not present as part of nature's realities but accessible only via supposition, assumption, and hypothesis – is part and parcel of the realm of mind-involving. For hypothesizing and its cognates (supposing, assuming, etc.) are one and all things that mind-endowed beings alone can do. Reality as such is factual and actualistic. Figuratively put, it talks the language of *is* and knows nothing of *may be*. It contains no mere possibilities and admits of no element of the merely hypothetical (over and above comprising mind-endowed beings who can project hypotheses). The only access route to the purely possible is that of supposition, assumption, and hypothesis – all of which call for mind-endowed beings to come upon the scene.

One salient argument for the mind-involvingness of mere possibilities runs as follows:

> To explain what a *mere* possibility is, one has to contrast it with a *real* possibility. A real possibility is one that can come to be realized through developed processes producing changes in real things (the possible oak tree that may emerge from this acorn). A mere possibility is one that cannot come about (be realized) through the machinations of the real; it can only be assumed (supposed, hypothesized). The concept of a mere possibility is accordingly mind invoking in the standard, explanatory ordered sense of that term. Possibilities as such can be either rear or mere; it is the mereness of a mere possibility that is conceptually mind-invoking.

To explicate what it is to be (merely) possible we thus need to involve the operation of minds. For the merely possible is not a part of the world's furnishings; it is something that can only be imagined, supposed, conjectured, hypothesized: that is, it is by its very nature something whose status is correlative with something that only minds can do.[3] The concept of the "merely possible" is accordingly an instance of a mind-invoking conception. Its adequate explanation requires a reference to the sort of thing that only minds can do.

Real possibilities are possibilities *de re*: "It is possible for Felix the cat to be on the mat." Hypothetical possibilities are *de dicto* possibilities: "It is possible for a cat (even a merely hypothetical cat) to be on the mat." Their status and condition in the explanatory nature of things is based on assumptions, suppositions. To explain adequately what mere possibility (in contrast to that of a real possibility) is all about, we must bring mental operations (assuming, supposing, hypothesizing) into the discussion. In view of this, hypothetical possibilities are inherently mind-involving. And this has important consequences for laws.

Consider the following exposition of one of conceptual idealism's key theses offered by one acute critic:

> Rescher's argument here rests on two points. First he maintains that fictional [i.e., non-existent, "merely hypothetical"] things must exist in some sense because they are said to be real possibilities; and then he argues that the only sense in which they could be said to exist is as the object of certain mental processes such as imagining or supposing.[4]

There is a substantial misunderstanding here. The position at issue is not the old Parmenides/Plato doctrine that even nonexistents must "exist" in some sense because they can be talked about (assumed, supposed, etc.). Our theory does *not* propose to establish the link between fictionality and mind-dependency by using existence as a middle term as per the argument:

- Fictional objects must have an existence of some sort.
- The only sort of existents that fictional objects can have is a mind-correlative one.

Therefore: Fictional objects have a mind-correlative existence.

This sort of *ontological* argumentation just is not at issue in the conceptual idealism envisioned here. For the fact is that the linkage between a fictionality (i.e., assumptive or suppositional or hypothetical status) and mind-involvingness is not enthymematically mediated but immediate, because fictionality (and its cognates) are inherently mind-invoking in their conceptual make-up. The point is that to make proper use of "merely possible objects" we must acknowledge them as fictions, and thereby characterized them in mind-referring terms of reference, seeing that fictions are, of necessity, the creatures of a mind.

But does not conceptual idealism's thesis that hypothetical possibilities are mind-invoking not run into difficulty because of a failure to distinguish between *what is supposed* (viz., the hypothetical possibility in question) and the *supposing* of it – this latter alone being mind-invoking. Are we not committing with reference to *supposition* the same old error that Berkeley committed with reference to *perception*, in refusing to distinguish what is perceived with the perceiving of it?

By no means! The cases are quite disanalogous. With existents, we are (*ex hypothesi*) at liberty to distinguish the thing at issue from the perceiving (contemplating, etc.) of it. Precisely because it exists: it is (*ex hypothesi*) there to so distinguished. But with nonexistents this prospect is all too plainly absent. Here Berkeley's doctrine goes awry: with *existents* we cannot maintain that to be is to be perceivable (let alone perceived) by us, simply because there can or might be actual things existing entirely outside our *perceptival* reach (gravity waves, for example). But with *nonexistents* the case is very different; here to be is to be exactly because the item itself, being nonexistent, is no more (but also no less) than an object of assumption supposition, and is constituted as the thing it is through the assuming or supposing of it. In this domain, to be is indeed to be conceivable.

There are, in theory, three distinct sorts of possibilities:

- real potentialities of actually existing things (if circumstances are favorable, yon acorn will grow into an oak tree);
- fictive possibilities for actually existing things (if yon acorn were a cat, it would have a backbone);
- mere possibilities (if a bolt of lightning were to hit here – not a repositioning of any of the actual ones, but a somehow additional one – then a thunderclap would sound here).[5]

But laws of nature apply to the sphere of possibility across the board (as the third case example shows). And it is crucial to their status as genuine laws that their application outruns the limits of the real into the realm of the merely hypothetical. (If a supposed "law" did not admit type-three implementation we would have to reject it as such.) If it is indeed a law of nature that mice have tails, then this will still hold good when I imagine (assume suppose) that there is a mouse in my pocket. For if it is indeed a (non-mutilated) *mouse* that is there, then a tail has to be there as well. But to explain and expound *this* absolutely essential feature of laws (viz., that they support counterfactuals relating to mere possibilities) we will have to bring assumptions (suppositions) into it – i.e., will have to make use of conceptions that relate quite explicitly to the operating of minds. Again, the thesis that "Oak trees are (lawfully) deciduous" – that is, "Oak trees *have to be* deciduous" in a sense that warrants "If that pine were an Oak, then it would be deciduous" – does not deal with actuality alone, but frames its claims about "the facts" in such a way as to bring in a realm of the hypothetically possible which by the very nature of the hypothetical, is inherently mind-involving.

To be sure, mere regularity calls for no more than a universal generalization of the type:

- All *X* is *Y*.

But genuine lawfulness goes well beyond this to call for stipulations of the form:

- All *X* is *Y*, and if any *z* that is not an *X* were an *X*, then *z* would be a *Y*.

The first generalization is simply an issue of existential fact: if it is in fact the case that *All (pure) mercury solidifies at - 150*°F, then this circumstance could continue operative even if one abstracted from all reference to minds, overt and covert alike. But a generalization of the second type – a truly lawful generalization – would fare differently. Its reference to *z*'s which are not *X*'s having to have certain characteristics if they were *X*'s involves claims outside the domain of ontological fact, claims of which sense can be made only by the use of mind-invoking conceptions of possibility. The mind's ability to use its concept-apparatus to project the specification of entities that, as it fully recognizes, have no standing in the world's scheme of things is one of its characteristic capacities – one that is crucial to operate the conception of lawfulness as we have it. The implicit projection of merely hypothetical possibilities is crucial to lawfulness and represents an element of intellectual artifice that is among the most obviously self-provided constructions the human mind affords us.

But, as another critic complains,[6] does not a law simply characterize *what does happen in this world* irrespective of any reference to mere possibilities? Not really. For while a law of nature characterizes what does happen in the world alright, it does so in terms of what *must* happen in the world – and thus does not effect its characterization without reference to possibilities. Laws are – and qua laws have to be – more than mere generalizations. For the crucial point is that laws support counterfactuals; they not only hold for the real things of this world but also for those mere possibilities which we project into it by means of suppositions, hypotheses and assumptions.

Lawfulness is the product of the well-founded imputation to empirical generalizations of nomic necessity and hypothetical force, whose nature is inherently (albeit implicitly) mind involving. Both pivotal factors, the essentially factual element of well-foundedness and the essentially decisional element of imputation, are necessary to laws. Well-founding is essential because the very spirit of the scientific enterprise demands reliance only upon *tested* generalizations that have a solid observational or experimental basis. But the element of imputation is also essential since – as emphasized above – we can only observe what *is*, i.e., forms parts of the realm of the actual, and not what corresponds to the modally necessary or the hypothetically possible. The nomic necessity and hypothetical force characteristic of lawfulness thus represent factors that a generalization cannot conceivably earn for itself on the basis of observational or experimental evidence alone: it has to be *endowed* with these factors. The basic feature of a lawful proposition is on this view not the *qualitative* aspect of being-a-law but the *relational* aspect of being-maintained-as-a-law. Lawfulness is in the final analysis a relational rather than an absolute and purely descriptive feature, with minds playing a key part in the relationship at issue, although – to be sure – a part only. For lawfulness involves possibility, and the gulf between the actual and the merely possible is one that can be bridged – and can *only* be bridged – by minds.[7]

There is a crucial difference between a claim's being a strictly descriptive truth about the real world as such and its being of the sort which – like a law – by its very meaning and purport goes beyond the real world. Descriptive statements about the real are embodied in assertions that limit themselves to features of the real. But statements about what is necessary or what is (merely) possible have to be construed in terms of reference to the domain of mere possibility as well. That something is necessary (holds in all [physically] possible worlds) or that it is merely possible (holds in some other [physically] possible – world) are claims which, by their very nature transcend the descriptive characterization of the real. The truth conditions of *such* facts do not turn simply on the descriptive constitution of the actual world, but invokes reference to the situation in "other possible worlds" as well – worlds to which our only mode of entry lies (in the very nature of the situation) only through the route of supposition, assumption, and hypothesis. For it is, clearly, only through thought – through the imagination as deployed in assumptions, suppositions, and hypotheses – that we can get cognitive access to other possible worlds.

It is necessary to face one rather immediate consequence of the doctrine of the mind-involvement of lawfulness: namely that *causality* will stand on exactly the same footing. To speak of causation is to invoke (however tacitly) the operation of causal laws. (That heating the water *caused* it to boil can only be maintained on a basis of some thesis of the general type of the causal law "water boils when heated to 100°C – at sea-level altitudes.") That any discussion of causation in the scientific context is coordinate with an at least implicit reference to causal laws is a part of general agreement among present-day philosophers of science. The very idea of a causal connection between events is deployed in a generalized way that envisages their relationship as governed by a law, a causal law. In this way, the argument inherent in our analysis of lawfulness as conceptually mind-dependent leads ineluctably to an acceptance of the traditional idealistic view that causality is mind-dependent. Thus relative to our standard conceptual scheme, the mentalistic resource of counterfact is required, *via* lawfulness, for our conceptualized apprehension of the realm of causally interrelated fact. In somewhat Kantian terms, our experience of the real as a network of causality involves and presupposes a wider realm governed by the category of possibility. To consider a relationship among actual occurrences within the space-time framework in terms of a *causal* linkage, is to regard it in a setting of powers and potentialities that inevitably ramify into the fact-transcending and mind-involving sphere of "what would happen if." If lawfulness is conceptually mind-involving, then so is causality.

5. Empirical Properties

What sorts of things are at issue when we ascribe properties to things? There are two principal prospects here: (1) The property at issue may be an empirical property, that is ascribed in virtue of the way in which things affect us – how they impact on our own senses or their instrumental extensions, or (2) the point, at issue is a dispositional one that serves to explain how things affect (or appear to affect) one another. But both roads to property ascription are mind-involving.

In the former, sensory (phenomenal) case, the situation with respect to mind-invocation is clear since only mind-endowed beings can have *experiences*. By an "*empirical* property" we shall here understand – quite in the literal sense of this term – one whose ascription to something reports the actual or potential result of experiential contact (be it direct or indirect) with this item. Such properties by their very nature indicate how experiencing agents – or the instrumentalities they employ – normally or standardly react to or interact with the thing to which the property is ascribed in observation and analogous processes. A manifest color predicate like "red" affords a paradigm example, because quite clearly any predicate that states how a thing standardly looks (or feels, smells, etc.) is patently mind-referential. But not only these "predicates of sense-observation," as we may call them, but also those of any mode of "observation" whatsoever – however complex the procedures and instrumentalities involved may be – will be of this sort. All such predicates that embody a claim as to how something is standardly-observed-to-be are clearly mind-referential, because of the inherently mind-referential nature of observation *per se*. Observation, after all, is in its essential nature a process of interaction in which a mind is necessarily one of the items involved: it is inherent in the very concept that "observing" is an *interaction* between a mind and some sector of reality. Thus even apart from the fact that complex modes of observations are (like measurement, as we have seen) dispositional and law-correlative – and so mind-involving – they are *also* mind-invoking because by their very nature they report the results of experiential interaction.

We turn now to *dispositional* properties that react to the capacities of things to elicit action characteristic responses to others. By their very nature as such, all such dispositional properties are law-correlative, and consequently mind-involving (in accordance with the prior argument to the effect that lawfulness and causality in general represent mind-involving conceptions). And so, while theoretical descriptive terms may, admittedly, fail to be directly observational (and so not mind-invoking in the dispositional manner at issue above), they are still inherently dispositional: since scientific predicates are inevitably applied only to characterize some lawfully *interactional* aspect of things.[8]

The mind-involvement of *dispositional* properties as inherently lawful suffices also to establish at once that of *all of our standard groupings of classification with respect to the lawful behavior of things*. And this has vast repercussions as regards the groupings of the classificatory system we standardly employ in our thought about nature. Thus the ascription to particulars of such classificatory features as "being made of iron," "being made of glass" or of "being a lion" or "being a maple" is just as dispositional (and hence mind-invoking) as being malleable, or fragile or being a conductor.[9] The conceptual mind-involvement of dispositions is the "thin end of the wedge" that can be used to establish the comparable status of all the empirical properties of the things of our ordinary, conceptual scheme that represent classifications linked essentially to lawful behavior.

6. Space and Time as Mind-Invoking

To have a position in space and time is ipso facto to be *locatable* in space-time – to be relatable to other things in terms of placement (above, below, right, left) and distance (nearer, more remote, etc.) and timing (now, earlier, later). But all *positioning* of things (be it in relating them to one another observationally or with reference to a coordinate-placement scheme of some sort) involves a mental process. The very idea of location and placement in particular, identifiable positions is conceptually mind-invoking. To locate something is to note or indicate its spatial or temporal placement in relation to something else, and this clearly involves identifying things and their positions, which, as we have seen, is something that is inexplicably bound up with the machinations of mind. Whenever a structure of *particular* (identified or identifiable) positions is at issue, mind involvingness comes upon the scene once again. If particular identification is geared in mind-invoking (as we have seen), then so is space-time positioning.

Moreover, spatial and temporal positioning by their very nature involve measurement. But observational mensuration of *all* sorts – the chronometric included – requires the specification of concrete individuals. For it is crucial that in any measuring process we deal at each given stage with one self-same measuring implement, with *the same* measuring rod or *the same* clock (or other periodic mechanism), reidentifying it and its parts as particulars. And, of course, such *reidentification* demands and indeed surpasses the demand for the identification of particulars. Thus if our case for the mind-involvement of particular identification has any merit, it works also to establish the mind-involvement of mensuration in general and chronometry in particular.

Our standard conception of what physical reality is all about pivots on the general idea of identifiable particulars located in space-time and interacting causally. And as the preceding deliberations suggest, all of the key elements of this conception (viz., particularity, causality, with spatio-temporality) are themselves conceptually mind-referring. Just this is the crux of our conceptual idealism.

7. The Problem of Circularity

It may seem plausible to offer the following sort of object against a conceptual idealism along the indicated lines:

> How can one sensibly maintain the mind dependency of matter as ordinarily conceived, when all the world recognizes that the operations of mind are based on the machinations of matter. (As Mark Twain asked: "When the body is drunk does the mind stay sober?") To be an idealist in the face of this recognition is surely to be involved in a vicious or at least vitiating circle.

However, this objection simply gets things wrong. There just is no question of any real conflict when the proper distinctions have been drawn, because altogether different sorts of dependencies or requirements are at issue in the two theses:

1. that mind is *causally* dependent upon (i.e., causally requires) matter, in that mental process demands *causally* or productively the physical workings of matter;
2. that matter (conceived of in the standard manner of material substance subject to physical law) is *explicatively* dependent upon (i.e., *conceptually* requires) mind, in that the conception of material processes involves *hermeneutically* or semantically the mentalistic working of mind.

We return once again the crucial distinction between the conceptual order with its essentially hermeneutic or *explicative* perspective upon the intellectual exposition of meanings, and the *causal* order with its *explanatory* perspective upon the causal rationalization of physical processes. In the hermeneutic framework of consideration, our concern is not with any facets of the causal explanation of intellectual processes, but upon understanding them from within, on their own terms. The issue is not one of *causal explanation* at all, but one of the *understanding* to be achieved through an analysis of the internal meaning-content of concepts and of the semantical *information* conveyed by statements in which they are operative. Because of the fundamental difference between these two perspectives, any conflict in the dependency relations to which they give rise is altogether harmless from the standpoint of actual inconsistency. The circle breaks because *different* modes of dependency are involved: we move from mind to matter in the conceptual order of understanding (of *rationes cognoscendi* or rather *concipiendi*) and from matter to mind in the ontological dependency order of causation (*rationes essendi*). Once all the due distinctions are duly heeded, any semblance of vicious circularity disappears. No doubt, this calls for a certain amount of care and subtlety – but then so do many issues of intellectual life, and why should things be easier in philosophy than elsewhere?

Conceptual idealism's line is that the sphere of matter is mind-involving in that the concept of concrete objects, located in space and time and interacting with one another, is shot through with mind-invoking conceptions. But this entire argument that mind is basic for matter is carried through in the conceptual order, based on the idea that a mind-invoking conception is one whose ultimate-analysis demands a reference to minds or their capabilities. The "mentalistic" or mind-referring sector of a conceptual framework is accordingly spanned by those of its conceptions whose rational exposition can only be accomplished in ultimately mind-invoking terms, that is, by means of concepts relating to characteristically mental capabilities. Conceptual idealism stakes no claims to priority of mind over matter in the of causal or ontological dependency; its position leaves open entirely the prospect that mind could hinge on matter in point of causation.

Accordingly, the conceptual idealist's thesis that one specific direction of dependence (viz., that of the physical upon that of the mental) is built into the view of reality at issue in our standard conceptual scheme must not be seen as conflicting with the debatable (but by no means thereby negligible) prospect that the scientific explanation of causal relationships might envision a reversal in the direction of dependence. Where different perspectives are involved, seemingly conflicting theses are perfectly compatible. (I can say without conflict that my car is economical in point of gas mileage and uneconomical in point of maintenance costs.)

But even if no vicious circle arises, do we not arrive at an equally vicious infinite regress that altogether precludes understanding. For is understandability not precluded from the outset if an adequate understanding of mind requires reference to its causal origins in matter and an adequate understanding of matter requires reference to its functional presuppositions of a mind-invoking sort? The answer is negative. Avoiding regress would be necessary only if one adopted an essentially linear model of understanding. But this is quite inappropriate in the case of *coor-*

dinated concepts such as the present instance of mind/matter or the simpler case of cause/effect. To say that we cannot fully understand the cause until we understand its effect, and that we cannot fully understand the effect until we understand the cause, is not to show that there is a vitiating regress with the result that we cannot understand either one. All it shows is that two such coordinated and interrelated concepts cannot be set out through a *sequential* explanation but must be grasped *together* in their systematic unity.

An analogy may be helpful at this point. Take a knife and its blade. If that object is to count as a knife, then that shiny thing attached to the handle must be a blade, but this thing cannot count as a blade unless the whole it comprises together with that handle is a knife. The two items stand in conceptually symbiotic apposition: *X* cannot be properly characterized as *X* unless it is duly related to *Y* and *Y* cannot be properly characterized as such unless it is duly related to *X*. We cannot pick up either end of the stick in separation from the other, but must grasp the whole in one fell swoop. Just such a cognitive *coordination* of mentalistic and materialistic concepts holds with respect to our present analysis of their mutual interdependencies. By maintaining due distinctions, any collapse into vicious circularity or vitiating regress can be avoided. The situation of biological symbiosis shows that reciprocal *causal* dependency need not be vicious, and indeed reciprocal conceptual dependency – ideational symbiosis – is certainly not. (Understanding is always systemic.)

Conceptual idealism is thus even compatible with a causal materialism that maintains matter to be basic to mind in the causal order. On the causal issues of the origins of mind, conceptualistic idealism is silent and so compatible with various theories – materialism itself not excluded. Conceptual idealism just is not a *genetic* theory regarding the causal mechanisms of the mind's processes or mode of origination; it is an *hermeneutic* theory regarding the nature of the conceptual mechanisms of the categories of understanding. It can thus coexist with *any* theory of mind that is articulated along strictly *causal* lines, be it a materialistic view that sees the causal origin of mind in matter of a Cartesian-style dualism of reciprocal influence or even an epi-phenomenalism.

The conceptual idealist accordingly has no vested interest in denying that mind and its functioning may be somehow causally emergent from the processes of matter. The doctrine's point is simply that our standard rational account of the world – its material sector specifically included – is given in terms of reference that are at bottom mind-involving. It is the analytical issue of *how* we think of the world, not the explanatory issue of its causal goings-on, that constitutes the focus of concern. This position does not need to be argued through an attack upon causal materialism: it is quite compatible with the idea that mental functioning has its material basis and causal origins in the realm of physical process.

Another critic has offered the following objection:

> His [Rescher's] argument [for the compatibility of materialism and conceptual idealism] is that, for the materialist, there is a *causal* dependence of mind on matter, whereas for the conceptual idealist there is a *conceptual* dependence of matter (reality-as-we-know-it) on mind. But I wonder if the issue is as clear as the simple recourse to a conceptual-causal distinction would suggest. If "matter" bears the same sense in both dependency claims, then the matter on which mind causally depends is itself conceptually infested with mind. Now, it is often said that, to make a genuinely causal claim, cause and effect must be independently describable. But if Rescher is right, then of course matter cannot be wholly described in terms independent of its supposed effect, mind.[10]

The fly in this objection's ointment lies in its construal of the "independent describability" of cause and effect. The validity of the causal thesis hinges on independent characterizability in the *causal* sense of independence. But any purely *conceptual* dependency will be quite immaterial as long as it does not intrude upon causal matters – as purely conceptual relationships would surely find it difficult to do. A purely conceptual role of mind in matter is accordingly quite perfectly innocuous as far as cause-effect considerations are concerned.

But what of an "identity theory of mind" that *identifies* mental processes with the operation of certain material configurations – namely our brains? Is our conceptual idealism not incompatible with such a theory? Not necessarily. It depends upon whether the identity is seen as a factual one (like the identity of the morning star with the evening star or that of the tallest man in the room with the poorest man in the room), or as a conceptually necessary one (like that of Smith's only brother with Smith's only male sibling). Our idealism will encounter no difficulties with a thesis of *contingent* identity. An incompatibility will arise only if the identity theory of the mental with the material is taken to obtain in *conceptual* terms, as holding the essentially concept-relative thesis that mentalistic talk is *eliminable*, in that it can be *translated without conceptually viable residue* into talk about the behavior of matter. Such a conceptually eliminative reductionism is indeed incompatible with a conceptual idealism. For if "mentalese" were analytically altogether reducible to materialistic discourse, then mind could not be conceptually basic to matter in the sense of our theory. But, of course, since our theory is based on an analysis of the *ordinary* conceptual scheme, this goes no further than to show that this ordinary scheme is incompatible with a conceptually reductive materialism, and *this* upshot is perhaps not surprising. (If we point out to the reductive materialist that he violates the ordinary conceptual scheme, he may well reply that he is only too ready to do so. And in taking this stance he is, to be sure, not inconsistent. But he does cut himself off from participation in those discussions that take place within the mentalistic framework of our standard conceptual scheme which is, after all, the frame of discussion in which the whole issue is posed in the first place.)

Conceptual idealism pivots on the fact that the conceptual instruments that the mind uses in forming its representations of reality are seen not only as *mind-made* (this point – though true – is altogether trivial), but that they are also in various respects *mind-patterned*, and so reflective of their mentalistic origin. The conceptual devices that mind deploys for constructing its view of reality are themselves heavily imbued – in the final analysis – with reference to the workings of mind. And it is this fact of the *analytical* fundamentality of mind-invoking (mentalese) conceptions – and not any considerations regarding its causal origins – that render mind basic to matter *in the conceptual order*.

One critic complains as follows:

> If, as Rescher seems to imply, it is just a matter of fact that we have experience of our own minds, then why should we find our factual experience of "mind-endowed individuals" more intelligible than that of non-mental particulars, or of "mind-inaugurated agency" more intelligible than that of non-mental causes?[11]

The difficulty that this objection poses is readily met. For, as Cicero noted (*Egomet mihi proximus sum*) and Descartes reemphasized, we are experientially closer to our own selves. A philosophy that respects the standpoint of empiricism in emphasizing the primacy of experience clearly does well to begin here with our own selves – and with the mental processes that are the stuff and substance of our thought-life.

8. An Elaboration of Consequence: What Conceptual Idealism Comes To

The most common objection to idealism in general centers on the issue of the mind-independence of the real. "Surely," so runs the objection, "things in nature would remain substantially unchanged if there were no minds. Had intelligent creatures never evolved on the earth, its mountains and valleys would nevertheless be much as they are, and the sun and moon remain substantially unaffected." This contention is perfectly plausible in one aspect, namely the *causal* one – which is just why causal idealism has its problems. But it is certainly not true *conceptually*. For the objection's exponent has to face the question of specifying *just exactly what* it is that would remain the same. "Surely roses would smell just as sweet in a mind-denuded world!" Well..., yes and no. Agreed – the absence of minds would not *change* roses. But rose fragrance and sweetness – and even the size and shape of roses – are all features whose character hinges on such mind-invoking operations as smelling, scanning, measuring, and the like. For something really to be a rose it must, unavoidably, have various capacities to evoke mental responses – it must admit of identification, specification, classification, and property attribution, and these, by their very nature, are all mental operations. And the situation is substantially the same with respect to those "objects" we call *the sun* and *the moon*. To be sure, the role of mind is here *hypothetical*. ("*If* certain interactions with duly constituted observers take place, *then* certain outcomes will be noted.") But the fact remains that nothing could be discriminated or characterized as a definite particular "object" in a context where the prospect of performing suitable mental operations (feeling, measuring, etc.) is not presupposed. All these matters in their very nature concern the relational issue of how roses impact on creatures of a certain sort (capable of smell, sight, etc.), and where one element of this relation is wholly absent (even in hypothesis), the whole cannot be present. Even in projecting a (hypothetically) mindless world we still make a tacit reference to the mental when we characterize a situation in mind-involving terms – in depicting things, for example, in terms of how minds would respond in interacting with them. And so a *conceptual* idealism remains in order.

On conceptual idealism's telling, there is something indelibly anthromorphic (man-patterned) or, more accurately, *noomorphic* (mind-patterned), in the way in which we think about the world. Against this view, someone may urge the objection that, "Surely, this does not show anything about the world itself, but rather about us." However, this objection loses sight of the fact that there is no clean and sharp separation between reality and the domain of thought, because our only possible route to cognitive contact with "the real world" is through the mediation of our conceptions about it – so that, for us, "the world" is inevitably "the world as we conceive of it." And in consequence, *how* we conceive of the world (namely in noormorphic terms) has to be seen as a fact not just about us, but about "the world" as well.

The conceptual idealism envisioned here is accordingly not a transcendental thesis about how matters must stand in general and in abstract. It does not maintain that *any possible* way of conceiving nature must proceed in mind-invoking terms of reference (difficult though it is for us to imagine how things could be otherwise). Its mission, rather, is to discern and characterize the role of mind-invocation operative in the *standard* conceptual framework that we in fact (*de facto*) use to recognize and interpret our experience.

To be sure, all that such a specifically *conceptual* idealism shows – and all that it is by its very nature in a position to show – is that reality *as we standardly conceive of it* is conceived of by us in mentalesque (i.e., mind-involving) terms of reference. It is a doctrine regarding our *conception* of reality rather than one about the *constitution* reality per se, about its descriptive rather than substantive make-up.

Conceptual idealism is not a doctrine about any and every *possible* conception of reality, but only one regarding what has here been characterized as "our standard conception of reality." Its strictures need not and will not apply to other possible conceptions of the real. Specifically, it does not apply to a contingent regularity view of the world that dispenses both with laws (and thus lawful processes): and with the disposition-demarkated particulars to whose conception lawfulness is indispensable. And while it is not easy to conceive in detail what such a kaleidoscopic reality that disposes both with particulars and with laws would be like, nevertheless, that does not render it impossible. All this must be conceded.

Is this concession damaging? Does it mean that conceptual idealism, with its focus in the ideational mechanisms in whose terms "we standardly think" of the real, really has no more than a sociological significance? The answer is clearly negative. For while how we *act* is simply a reflection of *sociological* matters, how we *think* of things is, clearly, an issue of deeper and more far-reaching *ontological* ramifications. There is, ultimately, no way of limiting the consequences of "how we think of things" to appertain merely and wholly to facts about *us* rather than facts about *them*.

Still, must not a genuine idealism ask for more? Must it not argue transcendentally that every possible conceptual scheme for exploiting experience to form a picture of objective reality must be mind-involving? Can it rest content with what is so *standardly* rather than *inevitably*? As one critic has objected:

> Rescher tries to handle the problem by, an appeal... to... the standard conceptual framework. But... the real and unavoidable problem is to determine the conditions of the possibility of *any* conceptual framework whatsoever.[12]

A splendid Kantian ambition, this – but very much misguided. For we surely cannot use the mechanism of our conceptual scheme to project from within the confines of that very conceptual scheme what the essential lineaments of other, different conceptual schemes must of necessity be. No state of science, no genre of art, no style of life or framework of thought can possibly manage to encompass all the rest. This sort of necessitarian thought-imperialism is just not in the cards. Our own cognitive position cannot at one and the same time be – as it must – just one position among others, and at the same time somehow *encompass* them by embracing their essential features. A state of knowledge or doctrinal position can never find the grail of self-transcendence – can never transmute itself into being more than something that has been to be presumed to be just one particular alternative among others. The envisioned quest for a self-transcendingly transcendental basis of "conditions under which alone conceptualization is possible" must accordingly be seen as a futile endeavor that is destined to failure from the very outset. (Why the implications of this fact are, for us, innocuous and in fact something positive rather than sceptically nihilistic will have to remain as the object of discussion for another day.[13])

That standard concept scheme of ours has to be taken at face value. No doubt, it is, in theory, simply one alternative among others, without any inevitable foothold in the very structure of human intellect. No doubt, its status is the product of natural and cultural evolution. Let all this be as it may. Still, *for us* the fact remains that this scheme is what we have and is all that we have. What matters in the end is that *this* alternative is *our* alternative. Our intellectual dependence on it is as absolute as our physical dependence on the air we breathe. For us there are no options. If this be "mere contingency," we have to make the most of it.

9. CONCLUSION

As long as one is prepared to draw the Kantian distinction between *actual reality* as such (i.e., reality as characterized descriptively by the *real* truth) and our putative or *empirical reality*, reality-as-we-think-of-it (i.e., reality as characterized by the presumptive truth of our *opinions* about it), then the present conceptualistic perspective has a useful and informative role to play. In particular, it precludes our embarking on the hopeless task of discussing "things in themselves" as a peculiar and mysterious *sort* of objects. They do not exist as a separate category that imposes on us the trouble of a separate consideration. To be sure, we *can* ask: "What can be said (appropriately and warrantedly) about the realm of existing things as we know it when we adopt the self-denying ordinance of abstaining from using any of the terms associated with the sorts of concepts we have characterized as mind invoking: individuality experientially accessible properties, causality, spatio-temporal features, etc.?" This is a meaningful question that demands an answer. And the answer is that we can say precious little. We can – perhaps even must – presume *that* there is existence in such a domain, but we cannot say *what* it is like. With all of our standard descriptive categories in abeyance, we are pretty well rendered mute with respect to the characterization of this realm – once we have presumed its existence and its existential essentiality to our familiar world. We are, in sum, in just about exactly the position that Kant is in with respect to the thing-in-itself.

The point, of course, is not what we can do for Kant, but what Kant can do for us – namely, to provide us with a conception essential to formulating a sensibly idealistic approach. For as Kant stressed, the reflexive "I think" is able everywhere to accompany our representations. We know full well that whatever information we have about the world's arrangements is information that has come to us through the workings of the mind – that our knowledge of the world's facts is inevitably and invariably mediated by mental operations. And in the final analysis we cannot avoid altogether facing the implication of the fact that, in our case, this mediation proceeds via concepts which, while inevitably mind-supplied, are also *de facto* implicitly mind-referring.

But is an idealism worth having if, like our *conceptual* idealism, its endorsement does not preclude one's being a causal materialist? The answer here depends very much on one's standard of what counts as worthwhile. If anti-materialism is one's prime standard of philosophical acceptability, then, clearly, such an idealism is not worth bothering about. But if one is sympathetic to idealism in its traditional sense of stressing the primacy of mind in the larger scheme of things, then conceptual idealism's emphasis on the centrality of mind is something one can regard as both illuminating and acceptable.

In sum, then, conceptual idealism as here understood maintains the mind-involving nature of concept scheme we standardly employ in the characterization of physical reality. But its stress is on the hermeneutic rather than causal role of mind. And its claim is staked not to the transcendental mind involvement of every possible concept scheme for characterizing the real, but the *de facto* mind-involvement of our *standard* scheme. Its necessities are not categorical but conditional, geared to how things have to be relative to and conditional upon certain basic *de facto* givens. It is the mundane realities of our ordinary concepts that this mode of idealism addresses, and not the ethereal region of necessary and inescapable "conditions under which alone." And this circumstance is something which, through deserving of remark, requires no apology. For, to quote Cicero once more, *ridiculum est quaerere quod habere non possumus.*[14,15]

NOTES

1. Some twenty years ago, in 1973 to be exact, my small book on this theme saw the light of print: *Conceptual Idealism* (Oxford, 1973). In the wake of its appearance, a number of reviews were published in the philosophical literature, and their perusal, as well as discussions with various colleagues, brought home to me how easy it is to misunderstand and misconstrue a position which, to its exponent, seems quite clear and straightforward. In due course, it came to strike me as useful to exploit the benefit of distant hindsight to try to restate the position more clearly and to try to safeguard it against various misunderstandings that have come to view. The present discussion is the result of this endeavor.
2. Phenomenal colors are mind-involving in that they have to do with the *looks* of things, and only mind-equipped beings can do perform looking in a literal sense; the language of looking is only figuratively applicable to mirrors or receiving instruments.
3. The point is not that only the merely possible be supposed (assumed, conjectured, etc.); it is that the merely possible can only be supposed – it cannot be positioned in reality's scheme of things in a mind-independent way.
4. Dennis Temple, "Conceptual Idealism and Unrealized Possibility," *Philosophy and Culture: Proceedings at the VXVI the World Congress of Philosophy* (Montreal: Congress Publications, 1988), pp. 44–50 (see p. 46).
5. The first two sorts of possibilities are unproblematically *de re*, but the third has to be cashed in through an ultimately *de dicto* reformulation.
6. Baruch Brody, "[Review of] Conceptual Idealism," *The Philosophical Review*, 84 (1975), 580–2 (see p. 581).
7. While laws are indeed man-made seeing that the law consists in what the law-statement that formulates it claim it to be, they are – of course – not thereby made as man *wants* them to be. In maintaining that the necessity and hypotheticality of lawfulness are matters of imputation, we

thus have no wish to suggest that the issue is one of indifferent conventions or arbitrary decisions.

8. Note that with *people* we might well draw a distinction between those properties they have "in themselves" (e.g., being intelligent or Spanish or music-loving) and those properties that merely indicate their impact upon others (e.g., being confidence-inspiring, repulsive, amusing, likeable, or the like). This clearly carries over into our talk about things (e.g., a statue can be bronze and Venus-representing on the one hand and lifelike and awe-inspiring upon the other). Now the point is that those scientific predicates that are not essentially classificatory (*felix leo*) or composition-indicative (carbon, gold) are for the most part of the sort that indicates an impact upon other things (conductor, acid, predator, etc.)
9. Cf. K. R. Popper, *Conjectures and Refutations* (New York, 1962), p. 387.
10. David-Hillel Ruben, "[Review of] *Conceptual Idealism*," *Mind*, 85 (1976), 138–40 (see p. 140).
11. Leon Pompa, "[Review of] *Conceptual Idealism*," *The Philosophical Quarterly*, 25 (1975), 85–7 (see p. 87).
12. Robert E. Innes, "[Review of] *Conceptual Idealism*," *Foundations of Language*, 14 (1976), 287–95 (see p. 294).
13. The issue is treated in the author's *The Strife of Systems* (Pittsburgh, 1985).
14. "It is absurd to ask for what we simply cannot have."
15. This essay is a revised version of a paper first published in *The Review of Metaphysics*, 44 (1991), 495–523.

CHAPTER 9

THE CONTRAST BETWEEN CAUSAL (EXPLANATORY) AND EXPERIENTIAL (NORMATIVE) UNDERSTANDING AND ITS RAMIFICATIONS

1. Two Approaches

As the medieval schoolmen already recognized, two rather different issues arise according as we concern ourselves with the relationships of the world's things to one another among themselves (*quoad se*) or with how they make an impact upon us *(quoad nos*). In our endeavors to come to cognitive terms with the world about us we correspondingly have at our disposal two very different modes of understanding, based on two very different cognitive perspectives – the explanatorily causal and the affectively experiential.

One cognitive approach is experiential, participatory, affective, phenomenological. It reflects the perspective of an intelligent agent whose frame of reference is based on those cognitive processes that inhere in its own mode of experiential interaction with things – those factors (real or supposed) of which it is conscious through its affective enmeshment in the world's scheme of things. Like any sort of understanding, the experiential mode is ideational and conceptual, but the ideas and concepts used here reflect the cognizer's own experience-geared position – the participatory standpoint of an inquirer's conscious cognitive involvement with the experiences that its interactive enmeshment in the world's processes provides. Here we have to do with interactively responsive point of view of an experiencing participant. Experiential understanding accordingly deals with the intentional and affectively evaluative aspect of things – with what things are:

- to be valued by us (∼ goodness);
- to be used for some purpose by us (∼ utility);
- to be understood as importing something or other for us (∼ significance).

Here we enter the specifically *normative* domain of meaning and value. We have to do with the axiological (= value related), teleological (= purpose related), and semiological (= meaning related) dimension of things in their bearing upon the concerns and interests of mind-endowed agents. Throughout, the factor of reflexivity and of us-correlativity inherent in the limitation "to be adjudged as such-and-such by/for us" is paramount. All of the characterizations at issue are forms of the normative stipulation that something can and should appropriately be treated by us in some particular way. If "primary" properties (in a more or less traditional sense) are features that things possess in themselves, independently of us, and "secondary" properties are features that their impact on us evokes in our thoughts, then experiential (participatory) understanding is inevitably concerned with those secondary properties that root in their interactions with us. The experiential understanding at issue here reflects our humanly affective involvement with things.

With causal explanation, by contrast, the concepts are those of a detached observer trying to get a purely cognitive grip on the ways of things irrespective of any involvement with them. This latter mode of understanding reflects the standpoint not of a concerned participant but of an intrigued spectator. The object here is not to grasp what it is like to respond experientially (to see or feel or undergo), but to *explain* – to tell a story that renders the things and processes at issue intelligible from an explanatory point of view. The causally explanatory understanding at issue here focuses upon impersonal factors and forces. It proceeds in terms of the standard mechanisms of causal or genetic explanation that reflect the lawful regularities of nature. The participatory element is absent. The mind is involved but the sentiments remain disengaged.

With explanatory understanding we function at the level of impersonal processes. (Q: How did he turn the light on? A: By closing the switch so as to complete the circuit which made the electric current...) With experiential understanding we function at the level of norms, values, and goals as they function in people's concerns. (Q: Why did he turn the light on? A: He was looking for his glasses and could not see properly in the darkened room.)

One of the most fundamental principles of epistemology is that different questions require different answers. And the two modes of understanding that concern us here address distinctly different sorts of issues. For explanatory understanding is oriented at *why* and *how* in a special sort of way. *Why* – in terms of the causal processes at work – did the occurrences at issue happen? *How* – in terms of those causal processes – did they come about? Here we proceed in an us-detached manner to explain the relevant phenomena in the mode of efficient causation, in terms of physical or developmental process, in terms of the forces, fields, physical, or biological mechanisms that are productively operative. Accordingly, in explanatory understanding we deal with how things interact with one another through the processes at work in nature's impersonal scheme of things.

But experiential understanding is something very different. Here we are dealing with how things affect us as sentient, conscious, intelligent beings. What now concerns us is the affective issue of how things bear upon our concerns and interests. The "things" involved may be the same as those that interest us explanatorily – sunrises, eclipses, flashes of lighting, etc. But the point of view from which we now consider them is altogether different. Different *dimensions* (as it were) of reality are at issue – different domains of discourse, different terms of reference, different concept spheres are at issue. One is a matter of productive processes – of causality in its broadest (classicism-transcending) sense. The other is a matter of personal reaction – of experiential phenomenology. We are dealing with conceptually distinct (autonomous and disjoint) domains of consideration.

To be sure, there is – no doubt about it! – always a causal story to be told about how people react by way of thought responses to imports of various sorts. But of course the causal origin of these reactions is one thing and their experiential nature another. (That's just what the present distinction is all about.)

Unquestionably, there is one point of contact – namely, ourselves. Our experiential reactions are also part of nature's operations – they too can be accounted for in the explanatory order of things. But that of course does not affect their inner character. We are creatures equipped with both body and mind, physical processes and thoughtful concerns. Both aspects exist and remain in place. Explanatorily we can consider our involvements in the world from the angle of causality – the angle of scientific explanation of the relation between thought states and body states. Experientially we can consider them from a phenomenological point of view of how it feels to participate in the various processes at issue. And there is of course no clash or conflict between impersonal process and personal import, between causation and significance.

The centrality from the standpoint of experiential understanding of the issue of how matters impress people and how people react to them in the realm of their own experience (the "by/for-us" aspect of the issues that figure in our experience) means that the artificial and man-made phenomena of literature, of artistic creation in general, and of the entire domain of our commerce with the imaginative/fictional and suppositional), is going to play a major formative role in this sphere. For us, as mind-equipped beings endowed with an imaginative capacity, the realm of speculative possibility acquires an importance and standing effectively comparable with the domain of what is can actually observe. (In *our* experience, imagination figures just about as prominently as observation.) Here, then, the arts acquire a status virtually equivalent with that of the natural sciences. Indeed, the perspective of a literature which explores the reach of how people respond to the vicissitudes of their experience is an eminently useful and effective entryway into the sphere of experiential understanding. The hermeneutical understanding of mind-geared meanings versus the causally explanatory understanding of matter-geared causes are fundamentally different intellectual processes coordinated with fundamentally different modes of intelligibility.

To be sure, there is no opposition and no conflict between the two orders of understanding. Each proceeds in its own conceptual dimension in terms of its own characteristic concepts. Indeed at the *conceptual* level there is no contact since each addresses different issues and deals with different aspects of the real. In moving from one angle of understanding to the other we effectively "change the subject." Two different domains of discourse, two different concept spheres are at issue.

In one important way, however, causally explanatory and normatively experiential understanding are alike. For the test of meaning and value is also in the end the test of experience – something that can and must be appraised by conducting a life that proceeds on its basis. Our knowledge of normativity – like that of causality – is ultimately a matter of inference from the data of experience. In *this* regard, both are alike. But there remains one crucial difference, because we deal with observational experience and *information* on the one hand, and with affective experience and *appreciation* on the other.

2. Historical Background

For over 100 years, philosophers have grapped with this dichotomy. It formed the core of Dilthey's distinction between the empathetic understanding (*verstehen*) that lies at the heart of the human sciences (*Geisteswissenschaften*) and the explanatory understanding (*erklären*) that lies at the heart of the causal explanations of the natural sciences (*Naturwissenschaften*). And after Dilthey, a whole generation of German philosophers (including Rickert, Windelband, Frischeisen-Köhler, and Münsterberg) elaborated with detailed various aspects of this issue. The idea that there is a fundamental difference of approach between the humanistic understanding of the biographer or political historian and the causally explanatory approach of the natural scientist. Human activity in its purposive and normative dimension is something that is not adequately understandable in terms of causal explanation with reference to general laws. And while causally explanatory science is value-neutral, normative inquiry is value-connected. The human sciences have an ineliminably axiological dimension: we cannot factor evaluation of them altogether.

The insistence that there is a profound difference between the hermeneutic perspective geared to the normative fabric of the humanistic disciplines and the causally explanatory perspective of the natural sciences represents a hard-won battle of the post-Dilthey generation. The next generation's positivists – with their rejection of axiology and their insistence on the methodological unity of the sciences (which in practice came down to saying that the mode of understanding of the physical sciences is all there properly is!) – is something that the following generation of post-World War II philosophers has had to struggle against anew in reaffirming the validity of modes of explanation geared to purposive rationality. But in its fundamental thrust the task at hand was a matter of moving not so much back to the future as forward to the past.

3. More on the Perspective of Science

The conceptual perspective of science is essentially causal. Be it natural, social, or psychological science that is at issue, the conceptual perspective at work is observational and explanatory rather than reactive and humanistically participatory. The data of natural science scientifically are those universals relating to interactions among observables than are accessible in principle to any and all observers, as contradistinguished from those that are in some degree person-geared and experience-reflective – accessible only to people of this or that particular background or experiential conditioning. Accordingly, science ignores the individualized, affective, and specifically humane dimension of human cognition: sympathy, empathy, feeling, insight, and "personal reaction." The phenomena it takes as data for its theory projection and theory testing are publicly accessible. Value appreciation – how things strike people by way of affective reaction and personal response is something that science leaves aside; it concentrates on the impersonally determinable – and preferably *measurable* – features of things. This quantitative orientation of our natural science means that the qualitative, affective, evaluative dimension of human cognition is sidelined. Our knowledge of the value dimension of experience – our recognition as such of these features of things in virtue of which we deem them beautiful or delightful or tragic – remains outside natural science's range of concern. Not experiential sensibility but theoretical/explanatory understanding is the crux.

It is the characterizing mission of natural science as a cognitive enterprise to enable us to explain the world's processes and eventuations, and to do this in such a way that we can test the adequacy of our explanations by means of effective technological intervention. But predictability geared description and causally-geared explanation are one sort of thing and interactive experiencing another. (We can *explain* how bats fly blind but cannot *experience* it.) And the crucial fact is that meanings and values arise in the experiential sector. To explain how people are involved in them is one thing, but actually to be involved with them is another.

Moreover, in natural science we investigate what is and what can be relative to the laws of nature. The more remote regions of possibility – the more imaginative realms of what might be and what might have been (let alone what *should* be!) – lie outside our purview in natural science, which, after all, does not deal with purely imaginative constructions and conjectured possibilities. Virtually by definition, natural science is oriented toward what actually exists in nature and does not deal with the speculative domain of imaginative projection nor the axiological domain of normative evaluation. Its concern is with the domain of actuality, the real, and that of the imaginatively hypothetical is outside its sphere. The artistic, imaginative, aspect of human creativity – the projection of strictly abstract form in art or music, for example – thus lies outside its range of concern. Poetry, drama, religion, proverbial wisdom, and so on all carry messages that cannot be conveyed within the medium of scientific discourse. The limits of science inhere in the limits of its cognitive mission and mandate: the "disinterested" depiction and rationalization of objective fact. The characteristic cognitive task of science is the *description* and *explanation* of the phenomena – the answering of our *how*? and *why*? questions about the causal workings of the world. Experientially geared normative questions of value, significance, legitimacy, and the like are simply "beside the point" of this project.

What matters for scientific observability is not specifically human sensory involvement but simply detectability – almost invariably by impersonal devices (meters and sensors) in which the role of human sensory observation is purely coincidental. With technological resources like radio telescopes or geiger counters and the like, science moves – with careful planning and deliberation – beyond the domain of human sensory observation. The type of "observation" at issue here is nowise specifically *experiential* in nature. Accordingly, science omits from its register of facts that are worth taking at face value the whole affective, emotional, feeling-oriented side of our cognitive life. Such matters appear within science as explanatory problems rather than as data for problem solving; they are seen as part of the problem rather than of the solution, as objects of study and explanation, not as usable materials. The sort of observation-transcending experience that lies at the basis of norms and values (affective receptivity, for example) remains off limits.

The "knowledge" of science is descriptive rather than normative, with the issue of evaluatively responsive appreciation – of what sorts of things are worthwhile – simply left aside. However value-laden the *pursuit* of scientific knowledge may be as regards the selection of problems and inquiry methods, on the side of its *content* (its substantive findings) science remains value-neutral, or perhaps better value-disconnected. Substantively considered, science is "value-free": its approach to the characterization of phenomena is simply devoid of the element of personalized evaluation. Natural science aims at mastery over nature – at controlling it both physically and intellectually through explanation and prediction. But it abstracts from the enterprise of *appreciating* nature or assessing our place within it. The interpretative, affective, appreciative, emotively evocative side of human knowledge that involves the normative responsive mechanisms through which we standardly understand other people and their productions falls outside the scope of science as insufficiently "objective."

And so, no matter how far we may push the scientific project forward along the physical, chemical, biological, and psychological fronts, there remain key issues about human concerns and works that will remain intractable by scientific means – not because science is impotent with respect to any of the issues that lie within its own range of concern, but because these matters lie outside this range. We shall always have experientially geared questions about ourselves and our place in this world's scheme of such a nature as to lie outside the explanatory problem mandate of science.

Owing to its exclusion of the normative issues of meaning and value, science approaches people as objects of study, as *things*, and not as *persons*. The sort of "knowledge" of a person in which it trades is not that of a friend or companion, based on concern inherent in mutual interaction, but the detached, "impersonal" knowledge of a doctor, biologist, or sociologist – the knowledge of observation rather than communion. The aspects of mutual recognition and interactive reciprocity are lost; and such cognitive modes as sympathy and empathy are put aside. The scientific mode of understanding takes the externalized route of causal explanation, and not the internalized route of affective interaction. The cognitive approach of science to the human understanding of humans deliberately puts aside that element of mutual recognition of ourselves as fellows and compeers characteristic of all genuinely human relationships among people. The work of science is explanation, not appreciation. It does not instruct me about matters of meaning and value. Such issues are not matters to be settled by scientific inquiry into nature's ways, which teaches us the why and wherefore of things, but does not instruct us about their worth.

Natural science, then, has its limitations. It is no be-all and end-all: it cannot answer all the questions that matter to us and has no monopoly on the business of understanding. But we are dealing here with the limitation of a range of concerns as such, and not with limitation of capacity within this range. It cannot be over-emphasized that what we are dealing with here is something that is not a defect or a shortcoming. The "limitations" at issue simply reflect mission-inherent boundaries imposed by the aims of the enterprise – the objectives that characterize science as the thing it is. Science has a determinate mission of its own in rationalizing the objective empirical facts; and here, as elsewhere, determination is negation: because science is a certain definite kind of enterprise, there are also things that science is not. These other cognitive ventures in understanding are not *alternatives* to science, because their cognitive orientation lies toward other directions. They are not different ways of doing the same job but look to doing altogether different sorts of jobs. And so, these other cognitive projects are not in a position to compete with science in its own domain; in fact, there is no question of competition, because diverse objections are at issue. Those who play different games are not in competition with each other.

The crucial fact is that the causally explanatory perspective of science upon the world's offerings is one among others. The fact that there are issues outside its domain is not a defect of natural science but an essential aspect of its nature as a particular enterprise with a mission of its own. It is no more a defect of science that it does not deal with "belles lettres" than it is a defect of dentistry that it does not deal with furniture repair. It is no deficiency of a screwdriver that it does not do the work of a hammer.

As the "idealist" tradition of German philosophy from Hegel to Heidegger[1] has always rightmindedly insisted, the quest for scientific knowledge is simply one human project among others. Reacting also against the Greek view that *epistêmê* is supreme, this tradition has replaced the old theological doctrine that *faith* transcends knowledge with the latter-day variant that *sensibility* transcends knowledge. Such a position has much to recommend it. (To be sure, its exponents have inclined to the rather tendentious view that experiential understanding is not only different from the explanatory understanding of science but somehow superior to it. But given the great differences of the issue mandate of these two endeavors, this is something like saying that literature is superior to music.)

Science does not have exclusive rights to "knowledge": its province is far narrower than that of human intellectualism in general. Even among the "modes of knowledge," science represents only one among others – the causal explanatory rather than the normatively experiential. It is geared to the use of theory to triangulate from process detection to answer our questions about how things work in the world. But there are other areas of human experience – areas wholly outside the province of science – in which we have a strong cognitive interest. Our concern with and need for understanding outruns the range of all specifically scientific concerns.

Natural science's mission-inherent feature of having a finite and determinate problem domain is something that the true friends of science are well-advised to keep in mind. Inflating the claims of science to the point where it is held to "have all the answers" about the condition of man, the meaning of life, or the objects of social polity is a dangerous policy. For it is always ill-advised to make exaggerated claims that go beyond what can be delivered. Such an inflated view of capacities invites scepticism and hostility in the wake of the disappointment of expectations that is its inevitable consequence. The upshot is bound to be a lamentable failure to secure proper credit for the inestimably great practical and intellectual benefits we derive from science within its proper domain.

4. Is Man a Machine?

But is the value of human life not decisively underminded by the unfolding progress of science – by the ongoing process of discovering that *Homo sapiens* is merely a machine – a mere part of the mechanisms of nature?

In 1748, Julien Offray de la Mettrie published in Leiden his startling treatise *L'Homme machine*, giving vivid expression to the naturalism that was part and parcel of the intensive dedication to "reason" that typified the intellectual ethos of the era that led up to and culminated in the French Revolution. However foolish de la Mettrie's view may have looked at a time when the most complicated machines were clockworks and windmills, it certainly seems far less so in the era of electronic "thinking machines" programmed to display "artificial intelligence." Perhaps then man is, after all, no more than a machine? Perhaps human life is meaningless as such.

The shock effect of the idea of man's being a machine lies in the very fallacious impression that this circumstance would somehow *change* what man in fact is. For we incline to think that this would mean that man would thereby be devalued, that human life would then cease to be significant. But how does this follow? If one accepts that our bodies consist of chemical elements – mainly water – whose market value is but a few dollars, does that have any implications for our value as *persons*? Why should it? The closer one looks, the less apparent the reason.

But if man were a machine, would that not mean that he has no soul? Well..., would it? What is "a soul" anyway? Clearly it is not some sort of physical component of one's body – like a kidney or a liver. Presumably a human soul is what it does – its being resides in a person's capacity to think, feel, aspire, love, and the like. Even if man were a machine, that would not mean that people lack souls. On the contrary – the consequence would simply be that machines can have souls. If man indeed were a machine, the net effect would be not to dehumanize man, but to produce a drastic chance in our understanding of the nature of machines by totally humanizing (at least) some of them.

People sometimes feel themselves threatened by the argument:

- Man is a machine.
- Machines cannot do X (cannot have feelings, have free will, etc.).

Therefore: Man does not do X (cannot have feelings, have free will, etc.).

But to maintain (2) is in effect to beg the question if (1) is asserted. All we can say is that *typical* machines do not behave in this way. But a human being, if indeed a machine, is certainly *not* a typical machine. The shock value of claiming that people are machines comes from the circumstance that it is all too easy to make the fallacious leap from the contention "Man is a machine" to the conclusion "Man is a *typical* machine." But of course this no more follows than from "Man is a *typical* animal."

Dictionaries generally define "machine" in such terms as a mechanical or electronically operated device designed for performing a predesignated task. Given *this* sort of specification, men and animals are obviously not machines. But this finding is not particularly interesting for the discussion because it rules out from the very start any prospect of considering the idea that one might be a machine. Such verbal force majeure apart, the question remains: is man a machine – albeit an organic, biological rather than mechanical, artifactual one?

To pursue this question intelligently would in the first instance require providing a pretty exact specification of just exactly what a machine is. And this is certainly not easy to come by. The breathtaking modern development in the capacities and complexities of "thinking machines" that makes it possible to contemplate man's being a machine without ludicrous incongruity have also cost us any secure intellectual grasp on just what it is to be a machine. Nobody has yet provided a clear and cogent explanation of just what "a machine" is in the current scheme of things. And in the absence of such an explanation, the question "Is man a machine?" poses an imponderable issue.

Even if it were to turn out in the end to be appropriate to categorize man as a machine, this eventuation would not necessarily dehumanize us. If man is indeed a machine, then he is certainly a very peculiar one – one that is organic, intelligent, capable of feeling, choosing, suffering, loving, etc. If man is a machine, then machines can do some pretty nonmachine-sounding things. The result would be to revolutionize not just our ordinary idea of *man* but our ordinary idea of *machine* as well.

Sometimes theorists try to bend the issue into other shapes. For example, might human beings be made artificially – somehow synthesized in a laboratory? But of course even if this could be done, it would not show we are machines. It would simply go to show that the range of things one can make in a laboratory includes not only chemicals and machines but also (perhaps more surprisingly) people.

To summarize. There is no good reason to think that we humans indeed are machines. But even if we were, we would certainly be machines of an unusual sort – namely, machines that care, machines endowed with free will, machines not designed and programmed by other agents but actually evolved in a way that puts them in charge of what they do themselves. Man's being a machine would not downgrade the value of human life, but would leave unaffected the value and stature of some (pretty nonstandard) machines.

But surely machines – unlike people – do not have a free will. Let us explore this issue.

5. Free Will

The issue of freedom of the will pivots on our clear sense that in many sorts of situations we are in control of what we do, that we are the authors of our acts, that what we consciously do can and often does issue from the deliberations we conduct in the light of our tastes, values inclinations, goals, and the like.

But are we indeed in charge of our actions? What of the dread prospect that we are mere automata in that our will is not free, and all our actions merely the result of the play of natural forces beyond our control? Surely minds are linked to bodies. (As Mark Twain asked: "When the body is drunk, does the mind stay sober?") And if bodies are indeed parts of the physical determinism of nature – and how could they fail to be? – then are not our thoughts and decisions also component of nature's causal commerce?[2]

People who deny free will generally proceed by way of contemplating a physical order of natural process that determines all the world's events, human actions included, through the operation of physical (and thus impersonal) causes. On this basis, they launch on the following plausible argumentation:

- Whatever people do falls within the scope of laws of nature.
- Actions falling within the scope of laws of nature are for this very reason necessitated in nature's impersonal order and are accordingly not free.

Therefore: Everything people do is determined by a causal necessitation operating independently of the wishes and choices of human agents.

On this telling, our actions are every bit as much necessitated by the impersonal forces of nature as is the downhill flow of water or the motion of a billiard ball. They only *seem* to us to be controlled by our choices and decisions – in actual fact, they are the product of an impersonal necessity that works in and through our brains and bodies.

This familiar line of reasoning is seriously flawed, however, in its projecting a facile opposition of causality and freedom. To have free will is, in essence, to be in a position to act from motives, goals, and values that one has – to have one's actions proceed from the "motivational" sectors of one's thinking via one's own decisions and choices. Freedom of action is something that lies in the experiential order of understanding. It is a matter of the "inner," thought-embedded determination of what one does, and not a matter of the severance of one's actions from causal determinations. A free act is one that is not free of *all* determination, but one that is free of determination *of a certain particular sort*, namely, from determiners that operate in detachment from one's own wants, preferences, wishes, and the like. The free agent "could have done otherwise," all right, but only if this agent's own wishes and choices and decisions had been different – which itself could have happened only if the agent's values, goals, and decisions had been different. No adequate explanation in the order of causal determination can factor the agent's own contributions out of it. The argument's crucial second premiss is accordingly false.

But is our capacity to implement our own agenda – to act on the basis of our own feelings, wishes, desires, preferences, etc. – really enough to render us free agents? For what if our motivation is itself set by "external" causal factors beyond our control? The reply is that everything now hinges on the *nature* of that "external" determination – on the sort of causal mechanism at issue. In particular, if this is a matter of manipulation by an external controller, thereby rendering our own putative decisions subject to the will of another, then the matter stands very ill for freedom. But if the causality in its control elements proceeds wholly from within, then freedom is undisturbed. Causality as such – one that operates on us only via the mediation of a specifically us-internal causality – is perfectly compatible with free agency.

Freedom of action is a matter of being in control of one's doings. It is a matter of autonomy, of independence of other *wills* – of hinging on one's own choices, preferences, wishes, etc., and being able to function independently of the choices, wishes, etc., of other agents. Freedom pivots on an agent's independence of other *agents* and their machinations – not of an independence of operative *causes*. After all, all of us – free agents included – are part of the world's fabric of lawful operation, and everything we do is to this extent conditioned and conceivably even "determined" by factors of nature and nurture over which we have no control. But *this* sort of "impersonal determination" has nothing to do with freedom of the will. What freedom precludes is not causality as such but a causality that bypasses the will of the agent – either because it involves no wills at all (i.e., no decisions, preferences, choices, etc.) or because it involves only the wills of *other* agents.

We thus arrive at the doctrine of *compatibilism* – the position that freedom of the will is compatible with a softened (duly internalized) determinism in the order of natural processes because "freedom" is not a matter of exemption from causal determinism but itself involves a determination of sorts, namely, one in which those "events" that are represented by decisions, choices, etc., and those "dispositions" that are involved in wishes, preferences, etc., are crucial for the adequate explanation of human actions. Freedom and determinism are thus not mutually exclusive contraries. In principle, an action can be at once determined *and* free, as long as the pattern of determination maintains its linkage to the wishes and choices of the agent: a causality mediated through the agents own volitional processes is perfectly compatible with freedom of the will. Free will is not at odds with determination because what it demands is just exactly determination by the will.

6. Science as the Destroyer of Values?

All the same, if the world is more or less as natural science nowadays depicts it, does this not deprive life of meaning and value? Does the world-picture of science not abolish any room for value in the world's scheme of things?

But how could it possibly do so? We come back at this point to our initial discussion of different modes of understanding. Natural science tells us a *causal* story; it describes the mechanics of how things work in the world. It does not – cannot – undermine the value of life because it does not address questions of value at all, seeing that it simply does not operate in the experiential order of understanding where value issues arise. The sound of a violin may be no more than the play of air waves produced by the vibration of dried cat's guts. But what has that got to do with the beauty or cacophony of the music? What matters for value is not the causal dimension of process, but the experiential dimension of product as it bears on the course of our lives. The issue of the meaningfulness of life is a value question. And this is something that science does not – and can not – affect in any negative way. Science is not and cannot be at odds with issues of meaning and value because these relate to issues that it does not address at all.

Many people sense a shudder of discomfort and dismay and even apprehension when science turns its cold, objective gaze upon man and his works – as though our deepest human values were somehow being put at risk by scientific understanding. But nothing could be more erroneous. To fear science as *antithetical* (rather than simply *indifferent*) to human values and interests is not only discomfiting, it is irrational and inappropriate, because based on a profound misunderstanding of what science is all about. The fact that a human being is an agglomeration of chemicals, a complex of flesh and bone evolved from creatures of the primeval slime, is no reason why he cannot also be a *friend*. Scientifically, the glow of a sunset no more than a shower of radiation. Does that prevent them from reaching heights of transcendent beauty? To speak of *conflict* here is to become mired in a confusion between diverse perspectives of consideration, between different levels of thought and inquiry, between the explanatory and experiential dimensions. Science is not *opposed* to these concerns, it is *irrelevant* to them; it simply ignores them, having other fish to fry.

In the final analysis, science has to leave the human world as it finds it – its work is to examine and explain the real, not to abrogate its dimension of meaning and value. In explaining how it is that people go about valuing things, science does not show that value is somehow unreal – any more than in explaining color vision, science shows that color is somehow unreal. Value attributions, like color attributions, are rooted in experience. And while science can presumably *explain* experience, it does not and indeed cannot show that it is somehow unreal or insignificant.

The theorist who maintains that science is the be-all and end-all – that what is not in science textbooks is not worth knowing – is an ideologist with a peculiar and distorted doctrine of his own. For him, science is no longer a particular sector of the cognitive enterprise but a self-sufficiently all-inclusive world-view. This is the doctrine not of *science* but of *scientism*. To take this stance is not to celebrate science but to distort it by casting the mantle of its authority over issues it was never meant to address.

Ludwig Wittgenstein wrote:

> We feel that even if *all possible* scientific questions be answered, the problems of life have still not been touched at all. Of course there is then no question left, and just this is the answer.[3]

The second statement here is as wrong as the first one is right. Its austere perspective pivots on the view that scientific issues are the only ones there are – that where no scientific question is at issue, nothing remains to be said, and that factual information is the end of the cognitive line. If this position is adopted, then questions relating to normative and evaluative issues of significance, meaning, and validity – questions relative to beauty or duty or justice, for example – can all be set at naught. Such a response does indeed resolve the "problems of life," but only by casting them away into the outer darkness. Unlike science itself, this scientific positivism is indeed antipathetic to human values.[4] But nothing within or about science demands such a dehumanization of our sensibilities. To take this stance is not to celebrate science but to distort it. It is to move from science to scientism, from science as inquiry to science as ideology.

Homo sapiens is a member not just of the *natural* but also of the specifically *human* order of things. There is more to our reality than science contemplates; in the harsh but stimulating school of life, we are set examinations involving problems for whose resolution our science courses by themselves do not equip us.

NOTES

1. See Martin Heidegger, *Being and Time*, ed. by J. Macquarrie and E. Robinson (London, 1962), sect. 32.
2. For an excellent discussion of the philosophical issues involved in the mind-body problem see Martin Carrier and Jürgen Mittelstrass, *Mind, Brain, Behavior* (Berlin, 1991).
3. Ludwig Wittgenstein, *Tractatus Logico-Philosophicus* (London, 1933), sect. 6.52.

4. As one acute writer has observed: "[Such a doctrine] is an attempt to consolidate science as a self-sufficient activity which exhausts all possible ways of appropriating the world intellectually. In this radical positivist view the realities of the world – which can, of course, be interpreted by natural science, but which are in addition an object of man's extreme curiosity, a source of fear or disgust, an occasion for commitment or rejection – if they are to be encompassed by reflection and expressed in words, can be reduced to their empirical properties. Suffering, death, ideological conflict, social classes, antithetical values of any kind – all are declared out of bounds, matters we can only be silent about, in obedience to the principle of verifiability. Positivism so understood is an act of escape from commitments, an escape masked as a definition of knowledge..." (Lasek Kolakowski, *The Alienation of Reason*, trans. by N. Guterman (Garden City, N.Y., 1968), p. 204).

CHAPTER 10

THE LIMITS OF COGNITIVE RELATIVISM

1. Cognitive Relativism

Claims to the truth of statements at the level of our scientific concerns involve an element of idealization. No doubt "the real truth" is one, but in this real and imperfect world of ours we have to accept the limitations of imperfection and face the fact that different inquirers living at different times and in different circumstances do have – and are bound to have – different ideas about the truth of things.

Relativism is the doctrine that people make their judgments by standards and criteria that have no inherent validity or cogency because their standing and status lies solely and wholly in the altogether "external" consideration of their acceptance by the group. The norms of different schools of thought differ, but those different norms are entirely on a par with one another.

In its general form, this relativist position has two prime components:

1. *Basis Diversity*: Judgments of the sort at issue – be it the true, or for that matter the good, the right, etc – are always made relative to a potentially variable basis: a norm, standard, criterion, or evaluative perspective regarding acceptability that change from one group to another.
2. *Basis Egalitarianism*. Any and every basis of normative appraisal – any and every such standard of assessment – is intrinsically as good (valid, appropriate) as any other.

Egalitarian relativism in general holds that it is no more rational to opt for one alternative than for any of its rivals.

Relativism thus understood does not deny that those who have a particular commitment (who belong to a particular school or tendency of thought) do indeed have a standard of judgment of some sort. But it insists that only custom speaks for that standard – that it is nothing more than just another contingent characteristic of the cognitive position of that particular group. It is all a matter of the parochial allegiances of the community – there is no larger, group-transcending "position of impersonal rationality" on whose basis one particular standard could reasonably and appropriately be defended as inherently superior to any other. The cognitive relativist, in particular, insists that no actual or possible group of inquirers whatsoever is in a privileged epistemic position. Every group's resolutions in this regard are final; each is its own final arbiter. There is no higher court of appeal – no inherently cogent basis of cognitive acceptability that has a compelling claim to validity.

Such cognitive relativism holds that there is no inherently good reasons for thinking some possibilities to be better candidates for claims to truth (more probable or plausible). The cognitive relativist does not go so far as to join the nihilistic sceptic in *rejecting* the reality of truth – he simply denies that there is no such thing as a good reason for applying this idea in any particular case.

Relativism insists that the difference between fact and opinion may be all very nice in theory but is simply unimplementable in practice. Whatever we may see as constituting "our knowledge" is simply a matter of opinion through and through. And one opinion is every bit as justified as any other. It's all just a matter of what people think. To believe anything else is no more than a recourse to the myth of the God's eye point of view – a point of view which, in the very nature of things cannot possibly be ours.

Over the last century, indifferentist relativism of this sort has gathered strength from various modern intellectual projects. As the sciences of man developed in the 19th Century – especially in historical and sociological studies – the idea increasingly gained acceptance that every culture and every era has its own characteristic fabric of thought and belief, each appropriate to and cogent for its own particular context. Historicist thinkers from Wilhelm Dilthey onwards have lent the aid and comfort of their authority to cultural relativism. And the aftermath of Darwinian biology reinforced this doctrinal tendency in giving currency to the idea that our human view of reality is formatively dependent upon our characteristically human cognitive endowments – as opposed to those of other possible sorts of intelligent creatures. Not only are the data about the world that we can acquire something that comes to us courtesy of the biological endowment of our senses, but the inferences we can draw from those data will be analogously dependent on the biological endowment of our minds. Various turn-of-the-century philosophers of otherwise very diverse theoretical orientations – ranging from Nietzsche and Vaihinger to Henri Bergson and Samuel Alexander – all drew heavily on Darwinian inspirations to support a syncretist perspectivism of one sort or another. Under such inspiration, William James wrote:

> Were we lobsters, or bees, it might be that our organization would have led to our using quite different modes from these [actual ones] of apprehending our experiences. It *might* be too (we cannot dogmatically deny this) that such categories, unimaginable by us today, would have proved on the whole as serviceable for handling our experiences mentally as those we actually use.[1]

Different cultures and different intellectual traditions, to say nothing of organically different sorts of creatures, will, so it is contended, describe and explain their experience – their world as they conceive it – in terms of concepts and categories of understanding substantially different from ours but in principle every bit as good. And anthropological and sociological investigations militated towards relativism on similar grounds. Moreover, psychologically inspired thinkers like F.S.C. Schiller and John Dewey were also impelled in very much the same direction.

In this way, investigations in various modern fields of inquiry have conspired to provide aid and comfort for the partisans of cognitive relativism.

2. Alternative Modes of Rationality?

Émile Durkheim was no doubt right in insisting that "all that constitutes reason, its principles and categories, is made (by particular societies operating) in the course of history."[2] But the fact that everyone's standards and criteria of rational cognition (etc.) are historically and culturally conditioned – our own of course included – certainly does not preclude their having a rationally binding stringency for those to whom they appertain, ourselves emphatically included. In conducting our affairs in this world – as in conducting our movements within it – we have no choice but to go on from where we are. If we are rational, then our standards and criteria of rationality presumably came to be ours precisely because we deem them to be (not necessarily the best possible in theory, but at any rate) the best available to us in practice on the basis of those considerations that we can conscientiously maintain.

But are the beliefs of primitive, pre-scientific cultures indeed less rational than ours? A resounding negative is maintained in Peter Winch's widely cited article on "Understanding a Primitive Society,"[3] which argues that Azande beliefs about witchcraft and oracles cannot be rejected as rationally inappropriate despite their clear violation of the evidential canons of modern Western scientific culture. But just here lies the problem. "Are those beliefs of theirs rational?" differs importantly from "Are those beliefs of theirs held rationally?" The answer you get depends on the question you ask. If we ask "Are those beliefs they hold rational?" we, of course, mean "*rational* on our understanding of the matter." And the answer here is clearly "*No*," seeing that by hypothesis our sort of rationality does not figure in *their* thinking at all. The fact that they (presumably) deem their beliefs somehow "justified" by some considerations or other that they see as appropriate is going to cut no ice in *our* deliberations regarding the cogency of those beliefs. ("Are they being 'rational' by *their* lights?" is one thing and "Are they being rational by *our* lights?" another.) When *we* ask about the rational acceptability of those beliefs we mean the issue to be considered from *our* point of view, and not from somebody else's!

The issues that arise at this juncture go back to the quarrel between E.E. Evans-Pritchard and Lucien Lévy-Bruhl. In his book on *Primitive Mentality*,[4] Lévy-Bruhl maintained that primitive people have a "pre-logical mentality." Against this view, Evans-Pritchard[5] argued that primitive people were perfectly "logical" all right, but simply used a logic *different* from ours. When, for example, the Nuer maintain that swamp light *is identical with* spirit, but deny that spirit *is identical with* swamp light, they are not being illogical, but simply have in view a logic of "identity" variant from that in vogue in Western cultures. But the obvious trouble with this sort of thing is that nothing apart from bafflement and confusion can result from translating Nuer talk by into *our* identity language if indeed it is the case that what is at issue in their thought and discourse nowise answers to our identity conception. Instead of translating the claim at issue as "Swamp light *is* identical with Spirit" and then going on to explain that "is identical with" does not really mean what it says because the ground rules that govern this idea are not applicable, an anthropologist would do well to reformulate or paraphrase (if need be) the claim at issue in such a way as to render intelligible what is actually going on. The fact that the Nuer have different (and to us strange-seeming) beliefs about "spirits" no more means that they have a logic different from ours than the fact that they communicate by drums mean that they have a telephone system different from ours.[6] To reemphasize: when we ask about logical acceptability it is logical acceptability by *our* lights that is at issue.

Anthropologists do sometimes say that a certain society has a conception of rationality that is different from ours. But that is literally nonsense. Those others can no more have a conception of rationality that addresses an object different form ours, than they can have a conception of iron that addresses an object different from ours, or a conception of elephants that addresses objects different from ours. If they are to conceive of *those* particular things at all, then their conception must substantially accord with ours. Iron objects are *by definition* what we take them to be; "elephant" is our word and *elephant* our conception. If you are not talking about *that*, then you are not talking about *elephants* at all. You have simply "changed the subject," and exited from the domain of the discussion. Similarly, if a conception of theirs (whatever it be) is not close to what *we* call rationality, then it just is not a conception of *rationality* – it does not address the topic that *we* are discussing when we put the theme of rationality on the agenda. Rationality as we ourselves see it is a matter of striving intelligently for appropriate resolutions – using *relevant* information and *cogent* principles of reasoning in the resolution of one's conjunctive and practical problems. If that is not what they are after, then it is not *rationality* that concerns them. The issues at stake in *our* deliberations have to be the issues as *we* construe them.

Of course, they may *think* that what we call pencils are chopsticks and use them as such. Or they may *think* books to be doorstops and use them as such. But, that does not mean that they conceive pencils or books differently from us, or that they have a different conception of pencils and books. "They take pencils to be something we do not (namely chopsticks)," is fine as a way of talking. But "They believe pencils to be chopsticks" is nonsense unless it is glossed as: "They believe these sorts of things called 'pencils' to be chopsticks." And when this happens, then they do not conceive of pencils differently from ourselves, they just do not conceive of pencils at all. They simply do not have the (one and only) available conception of pencils – namely, ours. In such cases, if they do not have *our* concept, then they just do not have *the* concept.

There is no difficulty with the idea that "They implement and *apply* the conception of rationality differently from ourselves." After all, we implement and apply the idea of a medication very differently from the ancient Greeks, using medications they never dreamt of . But the matter stands differently with the *conception* of a medication. This remains what it always has been: "a substance used as a remedy for an ailment." When one ceases to operate with what answers to that conception, then (*ex hypothesi*) one is no longer dealing with medications at all. The discussion has moved on to other topics. To have "a *different* conception" of pencils or elephants or rational actions is simply not to have that conception at all. If they do not have *our* conception of scissors, they do not have *a* conception of scissors, full stop. For, when we ask about their dealings with scissors it is our own conception that defines the terms of reference. If we recognize agents as rational at all, then we ourselves can make good rational sense of what they do! This is not because *we* are so talented and versatile, nor yet because the slogan "rationality is universal" gets it right. It is just because we could not and would not say that they are rational (could not and would not characterize the phenomena in this particular way) if we could not make rational sense of what they do.

So, it is literally nonsense to say "The *X*'s have a different conception of rationality from the one we have." For, if they do not have ours, they do not have any. It is, after all, rationality as we conceive of it that is at issue in this discussion of ours. Whatever analogue or functional equivalent there may be with which they are working, it just is not something that we, in our language, can call "a conception of rationality."

The pivotal fact lies in a "questioner's prerogative." Since the question "What is their mode of rationality?" is *ours*, so is the "rationality" that is at stake here. Thus, on the crucial issues – "What is rationality all about?", "What sorts of considerations characterize the rationality at issue?", "What is appropriately at stake?" – it is our own position that is determinative. When the questions are ours, the concepts that figure in them are ours as well. At this stage of establishing the constitutive ground rules of appropriateness for rationality, it is *our own position* that is decisive. Gestalt switches are certainly possible, but they are just that: unpredictable leaps. We do not reason our way into them by the use of existing standards – if it were so it would be an implementation of the old standard that is at issue and not, as per hypothesis, a Gestalt switch. They are only seen as rational *ex post facto*, from the vantage point of the *then* prevailing "established" standard – i.e., the new one.

Consider the contention:

> Surely there are no historically and culturally invariant principles of rationality. People's (altogether plausible) views about what is rational change with changes in place and time.

The response here is: yes and no. Of course, different people in different places and times conduct their "rational" affairs quite differently. But, at the level of basics, of first fundamentals, there is bound to be a uniformity. For, what all modes of "rationality" by hypothesis have in common is precisely this – that they all qualify as "modes of rationality" under *our* conception of the matter (which, after all, is what we're talking about). At this level of deliberation, "questioner's prerogative" prevails, and our own conception of the matter becomes determinative.

It is helpful to contemplate some analogies. There are many sorts of blades for knives. But, the fundamental principle that knives have blades at all does not depend on how people choose to make knives but on *our conception* of what a "knife" is. If the given objects, whatever they might be, do not have *blades* then they are not *knives*. It would, clearly, be the height of folly to go about in another culture asking people "Must knives have blades?"

The answer is a foregone conclusion. A negative response would not counter-indicate the thesis at issue, but would simply betoken a failure to comprehend.

Again, there are various quite different sorts of information that people take as evidence to substantiate a claim. But "deeper" principles like "Give more credence to that for which the evidence is stronger" or "In inquiry, endeavor to expand and extend the evidence for your claims" do not depend on the evidential practices of people, but on *our conception* of what evidence is all about. If people do not proceed in ways that conform to these evidential conceptions and principles of ours, then their practices – whatever they might be – are not *evidential* practices.

And so, in discussions about "alternative modes of rationality" we do indeed have a "higher standpoint" available to us – namely our own. And this is rationally justified by the consideration that no real alternative is open to us – we have to go on from where we are. Accordingly, while one must recognize the reality of alternative cognitive methodologies, one certainly need not see them as equally valid with one's own. "You have your standards and I have mine. There are alternatives." True enough. But this fact leaves me unaffected. For, I myself have no real choice about it: I *must* judge matters by my own lights. (Even if I turn to you as a consultant, I must ultimately appraise the acceptability of your recommendations on my own terms.)

3. A Foothold of One's Own: The Primacy of Our Own Standards of Rationality

Despite its fashionable pervasiveness, however, cognitive relativism has serious and indeed gravely debilitating defects. And in the main, these inhere in its commitment to basis egalitarianism.

The crucial question, however, is not "Are there indeed different norms and standards of rationality?", where the answer is an immediate and emphatic "yes." (Just as *autres temps, autres moeurs*, so also, other cultures, other standards.) Rather, the salient question is: "Are we well advised – perhaps even rationally obligated – to see all those various alternative norms and standards as equally appropriate, equally correct?" Must we adopt that Principle of Basis Egalitarianism: "All of the various standards of judgment have equivalent justifications. *Ours* is on an equal footing with *theirs* in point of acceptability. It is a matter of indifference which basis we adopt – each one is every bit as good (or poor) as the next"? And here the answer is an immediate and emphatic "no." There is no good and sufficient reason to see this principle as plausible. Basis indifferentism is daring and exciting – but also absurd. For, at this point we must turn relativism against itself by asking: "*Indifferent to whom*?" Certainly not to us! For, we have in place our own basis of rational judgement, and it speaks loud and clear on its own behalf. Nor yet by parity of reasoning is *ours* equally acceptable to *them*. From just what "angle of consideration" is it that claim to merit equivalency going to be made? Not from *ours* surely – for this, after all, is ours precisely because we deem it superior. And by parity of reasoning not from *theirs* either. (From God's? Well..., perhaps. But he of course is not a party to the discussion – and if he were, then what price relativism?)

Perhaps from the point of view of the universe all experiential perspectives are of equivalent merit; and perhaps they are equal before the World Spirit – or even God. But we ourselves cannot assume the prerogative of these mighty potencies. We humans can no more contemplate information with our minds without having a perspectival stance than we can contemplate material objects with our bodily eyes without having a perspectival stance. We ourselves do and must occupy a particular position, with particular kinds of concerns and particular practical and intellectual tools for dealing with them. And for us, there indeed is one particular set of standards for making such appraisals and adjudications, namely, our own – the one we actually accept in the conditions and circumstances of our existences. (And were we to trade this set of standards in for another, then of course that other one would automatically become ours.)

In approving positions we have, of course, no alternative to doing so from the perspective of our cognitive posture – our own cognitive position and point of view. (It wouldn't *be* our point of view if we didn't use it as such.) We cannot cogently maintain a posture of indifference. Each thinker, each school, is bound to take a strongly negative stand toward its competitors: belittling their concerns, deploring their standards, downgrading their ideals, disliking their presuppositions, scorning their contentions, and so on.

Cognitive (or evaluative or aesthetic, etc.) standard – perspectives do not come to us *ex nihilo*. From the rational point of view such standards themselves require validation. And this process – is itself something that is standard presupposing. For, of course, we cannot assess the adequacy of a standard-perspective in a vacuum, it must itself be supported with reference to standards of some sort. But in this world we are never totally bereft of such a basis: in the order of thought as in the world's physical order we always have a position of some sort. By the time one gets to the point of being able to think at all, there is always a background of available experience against which to form one's ideas. And just there is where one has to start. It is precisely because a standard-deploying certain position is appropriate *from where we ourselves actually stand* that makes this particular position of ours appropriate *for us*.

The salient point is that we are entitled – indeed, rationally constrained – to see our own criteriological basis of rational judgement as rationally superior to the available alternatives. If we did not take this stance – if we did not deem our cognitive posture effectively optimal – then we could not sensibly see ourselves as rationally justified in adopting it. We cannot responsibly think that this variant scheme as co-meritorious with our own, because (if rational) it is precisely because we deem that scheme superior that we have made it ours. It would, *ipso facto*, fail to be our real position – contrary to hypothesis.

In the final analysis, one can and should turn the weapons of relativism against itself. If indeed each group has its own standards against which there is no further, external, higher appeal, then we ourselves have no viable alternative but to see our conception of rationality as decisive for any judgments that we conscientiously make on these matters.

If we are going to be rational we must take – and have no responsible choice but to take – the stance that our own standards (of truth, value, and choice) are the appropriate ones. Be it in employing or in evaluating them, we ourselves must see our own standards as definitive because just exactly this is what it is for them to *be* our own standards – their being our standards *consists in our seeing them in this light.* To say that we are not entitled to view our standards as definitive could be to our having standards at all. And of course, someone who denies us this right – who says that we are not entitled to adopt those standards of ours – does no more than insist that it is by *his* standards (who else's) inappropriate for us to have these standards, and thus is simply pitting his standards against ours. We have to see our standards in an absolutistic light – as the uniquely right appropriately valid ones – because exactly this is what is at issue in their being our standards of authentic truth, value or whatever. To insist that we should view them with indifference is to deny us the prospects of having any standards at all. Commitment at this level is simply unavoidable. Our cognitive or evaluative perspective would not be our perspective did we not deem it rationally superior to others. We can cogently believe that "There are standards superior to mine", but we cannot cogently believe that "Such and such a particular (identified) standard is superior to mine", since seeing it as superior is (for natural beings) a compelling ground for adoption.

Anti-relativists often feel obliged to embark on a quixotic quest for "cognitive universals" at the level of substantive beliefs or cognitive procedures that all rational beings share in common. One can in principle contemplate two very different lines of approach here: (1) We find that certain types of creatures have somehow been predesignated as rational (by the World Spirit?), and we then inquire empirically into what it is that all of these predeterminedly rational beings have in common. Or again, (2) We make use of our conception of what rationality is to characterize certain types of creatures as rational, and *then* ask (analytically) what it is that all of them *must* have in common simply in virtue of qualifying under this conception of ours. Clearly, the second approach is the only practicable one; it makes no sense to try to implement the first (seeing that we simply have no way to get in touch with the World Spirit).

However, when we proceed in this second way, the only commonalities we can get out are the ones that we put in. We must ask what features beings must possess in virtue of qualifying as the sort of creature that *we ourselves are prepared to accept as answering to our conception of "rationality*." Clearly, this approach puts that conception of ours at the forefront as the determinative pivot point.

Recognizing that others see such matters differently from ourselves need not daunt us in attachment to our own views. It may give us "second thoughts" – may invite us to rethink – but this is not to move us to admit their standards but rather to make more careful and conscientious use of our own.

4. The Primacy of Our Own Position

When social scientists say that alien cultures have a different "rationality" from ourselves what they generally mean (strictly speaking) is (1) that they have different *objectives* (for example, that we seek to control and change our environment to suit our purposes, while they tend to reconstitute their purposes to suit their environment – to endeavour to come into "harmony" with nature), and/or (2) that they use *problem-solving techniques* which are different from ours (for example, that we employ empirical investigation, evidence, science, while they use divination, omens, or oracles). But, if they pursue different sorts of ends by different sorts of means they, perhaps, have a different thought style and a different intellectual ethos, but not a different *rationality*. The anthropologists' talk of different *rationality* is simply an overly dramatic (and also misleading) way of making a valid point – namely, that they do their intellectual problem-solving business in a way different from ours. But, those different processes of theirs do not mean that they have a different *rationality* any more then those blow-guns of theirs mean that they have a different *rifle*.

After all, it is *our conception* of rationality that fixes the "rules of the game" at issue when we pursue our deliberations about these matters. We have to play the rationality game by our ground rules because it is exactly those ground rules that define and determine what "the rationality game" is that is at issue on our deliberations. If we were not playing the game on this basis, it would not be the *rationality* game that we were engaged in – it would not be *rationality* that is the subject of our concern. It is the determinative role of our own rationality standards that makes them absolute for us.

Rationality is in this regard like communication. What communication is, is the same everywhere and for everyone – inherent in the nature of the concept that is at issue. But, of course, it is only normal and natural that different people in different places and times would transact their communicative business very differently, since what is effective in one context may fail to be so in another. Similarly, *what rationality is* is one thing (and one uniform thing from person to person within the framework of a meaningful discussion of the topic); but *what is rational* is something else again – something that is by no means uniform from person to person but variable with situation and circumstance.

To be sure, what makes our own conceptions authoritatively determinative in such matters is nothing special about us. Clearly, the Aristotelian cosmos is no longer with us – we are not the centre of the world. But we certainly and inevitably are at the centre *of our thought world*. Our inquiries have to be conducted within our frame of reference. We have to pose *our* questions in line with *our* ideas, to frame *our* perplexities by means of *our* concepts, to consider *our* issues in *our* terminology. If we ourselves are to classify someone as rational at all (and who else's attribution is now at issue?) then we must deem him qualified under the aegis of rationality *as we understand it*. If we ask about someone "Is she tall?," we are clearly asking about her height *as we conceive it*. What – if anything – she herself thinks about height is beside the point. And exactly the same situation holds with respect to rationality. A condition of "questioner's prerogative" prevails – it is the person who puts the issue on the table who sets the frame of reference for determining what that issue involves – it is, after all, *his* question. With all these questions about rationality, it is of course "rationality" *as we ourselves conceive of it* that is operative. The topic being ours, it is we who set the terms of reference for what is at issue. At this point "epistemic relativism" comes to a stop.

Paradoxically, it is precisely the inevitable relativization of *our* questions and concerns and puzzlements to *our* terms of reference that makes those particular terms of reference absolute in our own discussions. Being framed in our terminology, it is our terminology, that is decisive for the questions that *we* raise and the inquiries that *we* conduct. If *we* ask if *X* is being rational in believing (or doing or evaluating) a certain thing, then the issue is clearly one of its being rational on the basis of the conception of rationality as *we* understand it. The governing absoluteness of our conception inheres in "questioner's prerogative" – in the fact that the questions and issues we address in our deliberations about rationality are in fact our own and that, since the questions are ours, it is our conceptions that are determinative for what is at issue. In this regard, our commitment to our own cognitive position is (or should be) unalloyed. We ourselves are bound to see our own (rationally adopted) standards as superior to the available alternatives – and are, presumably, rationally entitled to do so on the basis of the cognitive values we ourselves endorse.

To qualify those of an alien culture as fully rational we must maintain both that they are conducting their inquiries intelligently by their own rules and also that in our sight these rules make good rational sense given their situation. It is ultimately "intelligent comportment" and "making sense" according to *our* standards of appraisal that makes what is at issue rationally invariant. The fact that *we* do (and must) apply *our* own idea of the matter is what makes for the universal element of rationality. What is universal about language use; namely, that to accredit another culture as rational at all is to accept it as being "rational" in *our* sense of the term – which may, to be sure, involve deciding whether their actions measure up to *their* standards. The absoluteness of (ideal) rationality is inherent in the very concept at issue.

5. AGAINST RELATIVISM

And so, while rationally appropriate knowledge claims and rationally appropriate actions and even criteria of appropriateness vary across times and cultures, the determinative principles of rationality do not. But, this interesting circumstance does not so much reflect a fact about different times and cultures as the fact that what *counts* as a "standard of rationality" at all is something that rests with us, because we are the arbiters of the conceptual makeup of an issue within the framework of our discussions of the matter. What sort of thing we ourselves understand by "rationality," becomes determinative for our own discussions of the matter. And this uniform conception of "what rationality is" suffices to establish and render uniform those top-level, metacriteriological standards by whose means each of us can judge the rationality of another's resolutions relative to that other's own basis of appraisal. For, those "deeper principles" of reason are inherent in the very conception of what is at issue. If you "violate" certain sorts of rules then – for merely *conceptual* reasons – you simply are not engaged in the evidential enterprise at all. The most basic principles of knifehood or evidence or rationality are "culture dependent" only in the sense that some cultures may not pursue a particular project (the cutting project, the evidence project, the rationality project) at all. It is not that they can pursue it in a different way – that they have learned how to make knives without blades, to evidentiate without grounds, or make rational deliberations without subscription to the fundamental principles we take as definitive of what rationality is all about. In such matters we do – inevitably and rightly as well – take our position to be pivotal.

These deliberations accordingly lead to a result that might be characterized as criteriological egocentrism. We can and indeed must see our own standards as optimal with respect to the available alternatives. If we did not so see them, then – insofar as we are rational – they would thereby cease to be our standards.

Indifferentist relativism insists that a choice between different bases of judgment must, if made at all, be based on extra-rational considerations – taste, custom, fashion, or the like. But rationality simply blocks the path to this destination in its demand that we attune our judgments to the structure of available experiences and its insistence that doing anything else would be *irrational.* Relativism thus has its limits. The implications of our own conception of rationality, truth, and inquiry are absolutely decisive for our deliberations. We ourselves must be the arbiters of tenability when the discussion at issue is one that we are conducting. And so, we cannot at once maintain our own rational commitments as such while yet ceasing to regard them as results at which all rational inquirers who proceed appropriately also ought to arrive given the same circumstances. In this sort of way, the claims of rationality are inherently universal and, if you will, absolute.[7]

"But is it not possible for someone to go out and get another normative standard?" It certainly is. But on what basis would one do this? You might *force* me to change standards. Or you can, perhaps, brainwash me. But you cannot *rationally persuade* me. For, rational persuasion at the normative level has to proceed in terms of norms that I accept and, by the norms I actually have, my present standards are bound to prevail, if I am rational in the first place.[8]

Thus, even while acknowledging that other judgements regarding matters of rationality may exist, we have no choice but to see our standards as appropriate for us. (In using someone else's with "no questions asked" we would, *ipso facto*, be making them ours!) To use another standard (categorically, not hypothetically) is to make it ours – to make it no longer *another* standard. Like "now" and "here," our standards follow us about no matter where we go. Of course, to see our own standards as the appropriate ones for us to use here and now is not to deny the prospect of a change of standards. But when this actually occurs, our stance toward the new standards is identical to our present stance towards the presently adopted ones. A commitment to the appropriateness of his present standards follows the rational man about like his own shadow. Relativity ends where charity begins – at home. For, our discourse is governed by *our* conceptions which are absolute at any rate *for us*. It is care for the *concepts* involved which chastens the impetus to relativism with absolutistic constraints.

To be sure, this criteriological egocentrism can and should be tempered by a posture of criteriological humility. The wisdom of hindsight and the school of bitter experience teaches us the chastening lesson that our cognitive standards – and the judgements we base on them – are by no means necessarily perfect. All the same, we have no real alternative to using our standards – to doing the very best we can with the means at our disposal. While we have to bear in mind the sobering thought that our best just may not be good enough, we are nevertheless bound to see the standards we have adopted in the pursuit of rationality as superior to the *available* alternatives and to regard ourselves as rationally entitled to do so here and now. (Future improvement "from within" can of course be envisioned along the above indicated lines.) To refrain from making this commitment is simply to opt out of the project of rational inquiry altogether. In the pursuit of rational cognition we must, as with any other pursuit, begin from where we are.

To acknowledge that other people hold views different from ours, and to concede the prospect that we may, even in the end, simply be unable to win them over by rational suasion, is emphatically not to accept an indifferentism to the effect that their views are just as valid or correct as ours are. To acquiesce in a situation where others think differently from oneself is neither to endorse their views not to abandon one's own. In many departments of life – in matters of politics, philosophy, art, morality and so on – we certainly do not take the position that the correctness of our own views is somehow undermined and their tenability compromised by the circumstance that others do not share them. And there is no good reason why we should see matters all that differently in matters of inquiry – or even evaluation.

6. The Arbitrament of Experience

A sensible pluralist will acknowledge not only that different people (groups or schools of thought) have different standards, but that they can do so appropriately given their differential emplacement in the cognitive scheme of things. But of course it does not follow from this that the pluralist need be disloyal to his own standards (any more than it follows from my acknowledging that your spouse to be appropriate *for you* constitutes on my part an act of disloyalty towards my own). To concede that certain standards make good rational sense for someone in your shoes is not to say that they make sense for one, who is differently situated.

The stance espoused throughout the present discussion has quite emphatically been that of a *perspectival rationalism* (or *contextualism*). Such a preferentialist position combines a pluralistic acknowledgement of distinct alternatives with a recognition that a group's choice among them is *not* rationally indifferent, but rather constrained by the probative indications of the *experience* that provides them with both the evidential basis and the evaluative criteria for effecting a rational choice.

Of course, people's experience differs. Even if we are pluralists and accept a wide variety of normative positions as being (abstractly speaking) available, still, if we have a doctrinal position at all – that is, if we are actually committed to solving our cognitive and normative problems – we have no serious alternative to seeing our own position as rationally superior. Faced with various possible answers to our questions, the sensible and appropriate course is clearly to figure out, as best we can, which one it is that deserves our endorsement. But, of course, there is – or should be – a good rational basis for effecting such a choice, namely normative perspective, whatever it be, that is based on the probative indications of our own experience. For there indeed is an objectifying impetus to avert our problem resolutions from becoming a mere matter of indifferent choice based on arbitrary preferences. And throughout the sphere of rational inquiry this objectifying impetus lies in the appropriate utilization of the lessons of experience. Empiricism is our appropriate and optimal policy. We have to go on from where *we* are and proceed on the basis of our experiential endowment. For us, a perspectival egalitarianism makes no sense when it comes to making our decisions regarding theoretical, practical, or evaluative matters.

Different individuals, different eras, different societies all have different bodies of experience. This being so, then whether discordant thinkers are at issue is not *their* perspectival accommodation of their experience just as valid for them as ours is for us? No doubt, the answer here has to be affirmative. But it must be followed by an immediate: "What of it?" The fact that others with different bodies of experience might resolve the matter differently is simply *irrelevant* to our own resolution of the issues. We have to go on from where *we* are and proceed on the basis of our perspective. For us, a perspectival egalitarianism makes no sense. Indifferentism is ruled out by the fact that it is experience that is the determinative factor and for us, the experience at issue is *our* experience and cannot be someone else's.

In matters of cognition as elsewhere, our normative orientations do not come to us *ex nihilo* but emerge from experience. And in this world we are never totally bereft of an experiential basis: in the order of thought as in the world's physical order we always have a position of some sort. By the time one gets to the point of being able to think at all, there is always a background of available experience against which to form one's ideas. And just there is where one has to start. It is precisely because a certain position is appropriate *from where we stand* that makes this particular position of ours appropriate for us. The posture that emerges from this way of approaching the issue is thus that of a contextualistic rationalism:

> Confronted with a pluralistic proliferation of alternative positions you have your acceptance-determination methodology, and I have mine. Yours leads you to endorse *P*; mine leads me to endorse not-*P*. Yours is just as valid for you (via your methodology of validating principles) as mine is for me. The situational differences of our contexts simply lead to different rational resolutions. And that 's the end of the matter.

The fact that the cognitive venture viewed as a whole incorporates other positions does nothing to render a firm and fervent commitment to one's own position somehow infeasible, let alone improper.

For us, our own experience (vicarious experience included) is something unique – and uniquely compelling. You, to be sure, are in the same position – your experience is compelling *for you* – but that's immaterial *to me*. "But isn't such an experiential absolutism just relativism by another name – is it not itself just a relativism of a particular sort – an experiential relativism?" Whatever relativity there may be is a relativization to evidence, so that relativism's characteristic element of indifference is lacking. It is just this, after all, that distinguishes indifferentist *relativism* from a rationalistic *contextualism*. And there is nothing *corrosive* about such a contribution: it does not dissolve any of our commitments. Its absolutism lies in the fact that, for us, our own experience is bound to be altogether compelling.

But does not a contextualistic pluralism put everyone's position on a par? Does it not underwrite the view that all the alternatives ultimately lie on the same level of acceptability? The question again is: Acceptable to whom? Rational inquiry as such maintains a certain olympian indifference – a noncommittal neutrality. However, this certainly does *not* mean that my position need be just as acceptable to you as to me. A sensible version of contextualistic pluralism will flatly refuse to put everyone's position on a par – save from the unachievable olympian point of view of the community at large which is, of course, by its very nature unavailable to any single individual. For each individual stands fully and decidedly committed to his own orientation on the basis of his or her own experience, so that there is no question of a relativistic *indifferentism* in acknowledging the pivotal role of a cognitive perspective. A pluralism of contextually underwritten cognitive positions does not lend to indifferentism precisely because a normative position is in the very nature of things something that a sensible person cannot view with indifference.

It makes no sense to take the line that all normative perspectives are equally acceptable, because where experiential bases of judgment are at issue, the pattern of our own experience is – for us at any rate – altogether decisive. After all, rationality requires that we attune our beliefs and evaluations to the overall pattern of our experience. For us, our own experience is rationally compelling. We could not (rationally) deviate from its dictates – and it would really make no sense for us to want to do so. We can no more separate from the indications of our own experience than we could separate ourselves from our own shadows. Such a position accordingly leaves no room for indifferentism. It pivots on the idea of *contextual* appropriateness – appropriateness in the context that is delimited and defined by the specific experiential circumstances of one's situation. Recognizing that pluralism prevails – that other standards are used by others – we nevertheless do and (appropriately) can deem our own standards of rational cognition as *appropriate* for ourselves. Even when conceding the prospect of someone's having another position, we cannot (rationally) see it as appropriate for ourselves.

7. The Impetus of Purpose: The Warranting of Our Position

Our own cognitive position may be but one among many, yet if we came by it aright, it is one whose adoption is appropriate for us. The pivotal fact of the matter is that cognitive rationality is something purposive – even though "theoretical" yet still it is, in a way, also "practical." For we require answers to our questions both to guide our actions in a difficult world and to respond to the information-requiring side of our nature.

Homo sapiens has evolved within nature to fill the ecological niche of an intelligent being. The demand for understanding, for a cognitive accommodation to one's environment, for "knowing one's way about: is one of the most fundamental requirements of the human condition. Humans are *Homo quaerens.* We have questions and (nay, *need*), answers. The need for information, for cognitive orientation in our environment, is as pressing a human need as that for food itself. We are rational animals and must feed our minds even as we must feed our bodies. In pursuing information, as in pursuing food, we have to settle for the best we can get at the time. We have questions and need answers – the best answers we can get here and now, regardless of their imperfections.

With us, the imperative to understanding is something altogether basic: things being as they are, we cannot function, let alone thrive, without knowledge of what goes on about us. The knowledge that orients our activities in this world is itself the most practical of things – a rational animal cannot feel at ease in situations of which it can make no cognitive sense. We have questions and require not just answers, but answers that cohere and fit together in an orderly way can alone satisfy a rational creature. This basic practical impetus to (coherent) information provides a fundamental imperative to cognitive intelligence.

We humans want and need our cognitive commitments to comprise an intelligible story, to give a comprehensive and coherent account of things. For us, cognitive satisfaction is unattainable on any other basis – the need for information, for knowledge to nourish the mind, is every bit as critical as the need for food to nourish the body. Cognitive vacuity or dissonance is as distressing to us as physical pain. Bafflement and ignorance – to give suspensions of judgement the somewhat harsher name that is their due – exact a substantial price from us. The quest for cognitive orientation in a difficult world represents a deeply practical requisite for us. That basic demand for information and understanding presses in upon us and we must do (and are pragmatically justified in doing) what is needed for its satisfaction. For us, cognition is the most practical of matters: Knowledge itself fulfils an acute practical need.

And so, seeing that knowledge can only be obtained from a cognitive perspective, the fact *that* we need such a perspective – some one or other – can be taken as a given. But this of course still leaves open the issue of *which one*?

The appropriate answer here is rather straightforward. The proper perspective for us is that which – in the existing circumstances

– most efficiently and effectively serves our purposes, which resolves our questions and furnishes our guidance with the most adequate and satisfactory way. While many alternatives doubtless exist, the one that deserves – nay, rationally requires – our endorsement is that which is best meets the needs of the existing situation as we face it. And this is going to be the perspective that is validated by our experience in that it most efficiently and effectively accommodates the concrete circumstances of our experiential situation.

"Facts are precisely what there is not," says Nietzsche, "only interpretations."[9] But this seemingly levelling perspective has its difficulties. For there are interpretations and interpretations – interpretations that resolve problems and interpretations that create difficulties. (If I interpret those grains of white sand as sugar and proceed accordingly I shall encounter unpleasant consequences.)

To be sure, we do not occupy the vantage point of a God's eye perspective. For us frail mortals, everything is a matter of judgment, of evidential availability, interpretation. But here we do have a guidance that is not of our making or wanting – the bitter school of actual experience. For where ideas are concerned humans may propose but nature disposes. Intelligent action may facilitate reaching our goals, but wishing and wanting is rarely of much help by itself.

What we have here, then, is an effectively *pragmatic* determination of the issue of the problem of perspectival selection and validation. There is an objectively optimal process of resolution through the adoption of that perspective which most efficiently and effectively accommodates the body of experience at our disposal.

To be sure, such a pragmatic/experiential validation of a particular cognitive perspective nowise *guarantees* the categorical correctness of the beliefs that ensue from it. All that it provides is that in following its strictures we will do as well as we can manage to in the circumstances. By allowing our beliefs by our experiences we do all that we can to assure their adequacy to our needs within the setting of the situation that confronts us. As indicated above, beyond that point – the point of having done all that we reasonably can – a certain element of methodological optimism is called for, a rationally appropriate hope that once we have done all we can these best efforts of ours will bear fruit.

8. Pluralism is Compatible With Commitment

But how, short of megalomania, can one take the stance that one's own view of what is rational is right – that it ought to be binding on everyone? How can I ask for this agreement between my position and that of "all sensible people"? Not, surely, because I seek to impose *my* standard on *them*, but because I do – or should! – endeavour to take account of their standards in the course of shaping my own. Co-ordination is achieved not because I insist on their conforming to me, nor yet because I supremely adjust my views to theirs, but rather because I have made every *reasonable* effort to make mine only that which (as best I can tell) ought to be everyone's. In the end, I can thus insist that they should use the same standard that I do because it is on this very basis of a commitment to impersonal cogency that I have made that standard my own in the first place. One's commitment to one's own rational standards is – or ought to be – produced not by megalomania but by objectivity. It is a matter of seeking – to realize a situation where I can reasonably expect that others will see matters in the same light as myself because I have found my light in the illumination of theirs.

Thus, nothing stands in the way of a realization that ours is not inherently and inevitably the best conceptual scheme – a kind of *ne plus ultra* – and that we can *nevertheless* stand rationally committed to it. Consider how cognitive progress happens. We can admit that the scientists of the future will have a better science, an ampler and more adequate understanding of the natural universe, and thus a better conceptual scheme – though, admittedly, we cannot anticipate just how this is to be so. We need not take the stance that our own conceptual scheme is somehow the last word. Our recognition that our scheme is imperfect, though correct and appropriate in the interests of realism, is, to be sure, of rather limited utility. A realization of the *en gros* deficiency of our conceptual machinery unhappily affords no help towards its emendation in matters of detail. A rational commitment to our position is just exactly a commitment to accepting its claims to be the proper basis for seeing rulings as proper and appropriate. It is a matter of doing the best we can in the cause of responsible truth estimation – of carrying on the business of rational inquiry as best we can in the circumstances in which the world's realities have emplaced us.

Nevertheless, this realization of the presumptive suboptimality of our own cognitive proceedings plays a most useful role. It affords a *regulative* conception that preempts any claim to dogmatic finality, even if not a *constitutive* one that puts substantively informative data at our disposal. The presumption that there are other and perhaps better ventures of inquiry than ours is eminently salutary in blocking the path of the deplorably egocentric view that we ourselves somehow stand at the cognitive center of things, occupying that pivotal position about which all else revolves. And here – "internally" from our own cognitive posture – we ourselves simply cannot consider other positions as genuinely on a par with our own in point of merit.

An experiential pluralism of cognitive positions is thus no impediment to a commitment to the pursuit of truth. There is no reason that the mere existence of different views and positions should leave us immobilized like the ass of Buridan between the alternatives. Nor are we left with the gray emptiness of equalitarianism that looks to all sides with neutrality and uncommitted indifference. In important matters like inquiry (or evaluation, etc.)

we cannot – and in good conscience should not – bring ourselves to view disagreement in the light of a "mere divergence of opinion."

It is, in the eyes of some, a disadvantage of pluralism that it supposedly undermines one's commitment to one's own position. But this is simply fallacious. There is no good reason why a recognition that others, circumstanced as they are, are inclined – and indeed rationally entitled *in their circumstances* – to hold a position of variance with ours should be construed to mean that we, circumstanced as we are, need feel any rational obligation to abandon our position. Once we have done our rational best to substantiate our position, the mere existence of alternatives need give us no pause.

Jean Paul Sartre deplored the attempt to secure rationally validated knowledge, which he saw as a way of avoiding responsibility for *making* something of oneself, for "choosing one's own project," seeing that the real truth is not something one can make up as one goes along but is something one regards as entitled to one's recognition (to a subordination of sorts on one's part). But this view turns the matter topsy-turvy. Not the pursuit of truth but its *abandonment* represents a failure of nerve and a crisis of confidence. The avoidance of responsibility lies in an indifferentism that sees merit everywhere and validity nowhere (or vice versa), thereby relieving us of any and all duty to investigate the issues in a serious, workmanlike way.

Admittedly, there are cognitive postures different from ours – different sorts of standards altogether. But what does that mean *for us*? What are *we* to do about it? Several stances towards those various bases are in theory open to us:

1. accept none: reject ours
2. accept one: retain ours
3. accept several: conjoin others with ours
4. "rise above the conflict": say "a plague on all your houses" and take the path of idealization invoking the "ideal observer,"

the "wise man" of the Stoics, the "ideally rational agent" of the economists, or the like

The first option is mere petulance – a matter of stalking off in "fox-and-grapes" fashion because we cannot have it all our own preferred way. The third option is infeasible: different bases do not combine, they make mutually incompatible demands, and in *conjoining* them we will not get something more comprehensive and complete – we will get a mess. The fourth option is utopian and unrealistic. We have no way to get there from here. Only the second alternative makes sense: to have the courage of our convictions and stand by our own guns. In *evaluating* contentions and positions (of any sort) we just have no plausible alternative to doing so from the perspective of our cognitive values – our own conscientiously adopted cognitive point of view.

To obtain informative guidance from inquiry, it is not enough to *contemplate* cognitive standards – be it as historical actualities or as theoretical possibilities. We must actually *commit* ourselves to one. We can only get viable answers to substantive questions if we do our inquiring in the doctrinalist manner – only if we are willing to "stick our necks out" and take a position that endorses some answers and rejects others, to be sure, in a principled way that is in line with standards for whose due validation we have made appropriate provisions. A perspectively grounded position is good enough for sensible individuals precisely because that position-determinative perspective of ours is by hypothesis our perspective.

9. On Pursuing "The Truth"

Regrettably, many people have – under the influence of relativism – simply given up on the truth. The very idea of "the truth" is of small interest to various theorists nowadays. Heidegger, for one, regarded those so-called absolute truths as no more than "remnants of Christian theology in the problem field of philosophy."[10] Of themselves, truth and falsity, correctness and incorrectness, adequacy and inadequacy, reason and unreason, sense and nonsense – approached as issues of logic, semantic theory, or epistemological explication – simply do not interest the hermeneuticist. He wants to know what role the *ideas* about these issues have in the sphere of authentic human experience; he does not ask about what these ideas *mean* but about what people *do* with them. Truth as such is something he is eager to abandon.

And his is not alone. In stressing the pluralism of philosophizing, William James wrote:

> *The* Truth: what a perfect idol of the rationalistic mind! I read in an old letter – from a gifted friend who died too young – these words: "In everything, in science, art, morals, and religion, there *must* be one system that is right and *every* other wrong." How characteristic of the enthusiasm of a certain stage of young! At twenty-one we rise to such a challenge and expect to find the system. It never occurs to most of us even later that the question "What is *the* Truth?" is no real question (being irrelative to all conditions) and that the whole notion of *the* truth is an abstraction from the fact of truths in the plural.[11]

Inspired by James, various contemporary pragmatists are quite prepared to abandon concern for truth.

But this reaction is gravely misguided. Epistemological pluralism has no substantive consequences for the nature of truth as such. The fact that "our truth," the truth as we see it, is not necessarily that of others – that it is no more than the best *estimate* of the real truth that we ourselves are able to make – should not disillusion us in our inquiries and should not discourage us in "the pursuit of truth." In inquiry as in other departments of human endeavor we are well advised simply to do the best we can. Realizing that there are no guarantees we have little sensible choice – *pro tem* at least – but to deem the best we can do as good enough.

Recognizing that others see matters differently in contexts of inquiry need not daunt us in attachment to our own views of the matter. There is clearly no conflict between our commitment to the truth as we see it and a recognition that the adoption of a variant probative perspective leads others to see the truth differently.

Given that we ourselves occupy our perspective, we are bound to see *our* truth as *the* truth. But we nevertheless can and do recognize that others who operate in different times and conditions see matters in a different light. The circumstance that different people see something differently does not destroy or degrade the thing as such.

The fact is that relativism misconstrues the significance of unavailable consensus. For, after all, consensus as such is neither a means to nor an effect of people's commitment to rational cogency. We have to come to terms with the epistemic realities, which include:

- the diversity in people's experiences and cognitive situations;
- the variation of "available data";
- an underdetermination of facts by data;
- the variability of people's cognitive values (evidential security, simplicity, etc.);
- the variation of cognitive methodology and the epistemic "state-of-the-art".

Such factors make for a difference in the beliefs, judgments, and evaluations even of otherwise "perfectly rational" people. Short of a biological and situational cloning that equips every inquirer with exactly the same cognitive basis for the formation of opinion, it is clear that consensus is unrealizable.

Rationality can be counted upon to lead to *consensus* only in situations of uniform experience – which are, obviously, also in general not realized. In the circumstances in which we labor in this world, consensuality is neither a requisite for nor a consequence of rationality in the conduct of inquiry. Nor is it a practicable goal. All of our factual claims are made in a setting of place and time, in the context of cultural and observational states of the art. But their being provisional and contextual claims to truth does not metamorphose them into claim to conditional and contextual truth. The flaw of relativism is that it conflates the *status* of our truth claims with their *substance*.

We have no choice but to pursue *the* truth by way of cultivating *our* truth; we have no direct access to truth unmediated by the epistemological resources of rational inquiry. And, given the ground rules of rational inquiry, this means that one's view of the truth is bound to be linked to one's cognitive situation. To say that this is not good enough and to give up on our truth – to declare petulently that if we cannot have *the* absolute, capital-T Truth that of its very nature constrains everyone's allegiance, then we will not accept anything at all – is automatically to get nothing and to abandon the pursuit of truth as such. It is foolishness to reject an orientation-bound position as not worth having in a domain where a position is only to be had on this basis. The only truth claims worth staking are those which can be seen as rational on the standards that we ourselves endorse.

10. THE ACHILLES HEEL OF RELATIVISM

But are we really entitled – rationally entitled – to place such reliance on our own standards? Assuming that we indeed have done what reason – as best we can understand it – demands for their substantiation, the answer is and has to be an emphatic *yes*. Inquirers should not and need not be intimidated by the fact of disagreement – it makes perfectly good sense for a person to do the best possible towards securing evidentiated beliefs and justifiable choices without undue worry about whether others agree or disagree.

There is nothing admirable in relativism's inclination to noncommittal detachment with its concomitant reluctance to trust one's personal judgment in matters of human significance. Relativism reflects a regrettable unwillingless to take intellectual responsibility – to say: "I've investigated the matter as best I can, and this is the result at which I've arrived. Here I stand, I can do no other. If you wish to stand with me, then welcome to you; if not, then please show me how my position is untenable." Relativism takes recourse to an uncritical open-mindedness that comes down to empty-mindedness. In matters of rational inquiry, as in politics and religion, one does well to prefer someone who has views and sticks by them to those who reject the whole project or (equally wrongly) try to ride off in every direction at once.

The decisive weakness of a cognitive relativism with its commitment to basis egalitarianism is matters of cognition lies deep in the nature of the human condition. It emerges as follows: The characteristic stance of relativism is to insist on the *rational indifference* of alternatives. Be it contentions, beliefs, doctrines, practices, customs or whatever that is at issue, the relativist insist that "it just doesn't matter" in point of rationality. People are led to adopt one alternative over another by *extra-rational* considerations (custom, habituation, fashion, or whatever); from the rational point of view there is nothing to choose – all the alternatives stand on the same footing. The fatal flaw of this position roots in the fact that our claims, beliefs, doctrines, practices, customs, all belong to identifiable departments of purposive human endeavor – identifiable domains, disciplines, and the like. For all (or virtually all) human enterprises are at bottom teleological – conducted with some sort of end or objective in view. In particular, the cognitive enterprise of rational inquiry has the mission of providing implementable information about our natural and artificial environments – information that we can use to orient ourselves cognitively and practically in the world. After all, explanation, prediction, and effective intervention constitute definitive enterprise-characterizing goals of science. (And the moral enterprise is also purposive, its mission being to define, teach and encourage modes of action that bring the behavior of individuals into alignment with the best overall interests of the group.) Human endeavors in general have an inherent teleology seeing that *Homo sapiens* is a goal-pursuing creature. Now in this context, the crucial fact is that some claims, beliefs, doctrines, practices, customs, etc. are bound to serve the purposes of their domain better than others. For it is pretty much inevitable that in any goal-oriented enterprise, some alternative ways of proceeding serve better than others with respect to the relevant range of purpose, proving themselves more efficient and effective in point of goal realization. And in the teleological contexts *thereby* establish themselves as rationally appropriate with respect to the issues. It lies in the nature of the thing that the quintessentially rational thing to do – in inquiry as elsewhere – is to give precedence and priority to those alternatives that are more effective with respect to the range of purposes at issue.

There is no doubt that such a position qualifies as a version of "pragmatism." But it is crucially important to note that it is *not* a version of practicalism. It stands committed to the primacy of purpose, but it certainly does not endorse the idea that the only possible (or only valid) sort of human purpose is that of the type traditionally characterized as "practical" – i.e., one that is geared to the physical, and "material" well-being of people. Purposive enterprises are as diverse and varied as the whole spectrum of legitimate human purpose, and such purposes can relate not only to the "material" but also to the "spiritual" side of people (their knowledge, artistic sensibility, social dispositions, etc.). To distinguish the position from the classical range of specifically *practical* purposes such a pragmatism might perhaps better be characterized as *functionalism*.

Such a functionalist perspective is decisive in its impetus against relativism. For relativism with its commitment to basic egalitarianism is flatly indifferentistic. Presented with alternatives of the sort at issue (cognitive, moral, etc.), the relativist insists that at bottom it just doesn't matter – at any rate as far as the rationality of the issue is concerned. But once we see the issues in a purposive perspective, this line just doesn't work. In a purposive context, alternatives are not in general portrayable as rationally indifferent. Rationality not only permits but demands giving preference to purposively effectual alternatives over the rest (at any rate as long as other things are anything like equal). It is quintessentially rational to prefer what works.

To the relativist one must grant *pluralism* – the lack of uniformity and the availability of context – variable alternatives. But this concession does not "give away the store" because it does not authorize irrationalistic indifferentism. On the contrary, two higher level meta-norms – to wit, conformity to one's standards and conduciveness to one's goals – operate so as to assure a purchase hold for rational appropriateness at the procedural level (even though they do not otherwise engender substantive uniformity).

As this line of reflection indicates, relativism stubs its toe against the pervasively purposive nature of the human situation, the fact that our proceedings – be it in inquiry, interpersonal interaction or whatever – fall within the scope of purposive ventures that have an end in view. This teleological aspect provides a basis for rationality in a way that puts relativism into a thoroughly problematic and dubious light.[12]

NOTES

1. *Pragmatism* (New York, 1907), p. 171. The basic line of thought goes back to the ancient sceptics. Compare Sextus Empiricus, *Outlines of Pyrrhonism*, I, 54, 59–60, 97, *et passim.*
2. This view has become axiomatic for the entire "sociology of knowledge."
3. Peter Winch, "Understanding a Primitive Society," *American Philosophical Quarterly*, 1 (1964) 307–24; reprinted in B.R. Wilson (ed.), *Rationality* (Oxford, 1970).
4. Lévy-Bruhl, *Primitive Mentality* (London, 1923; first pub. in French, Paris, 1921).
5. *Witchcraft, Oracles and Magic Among the Azandi* (Oxford, 1937); *Nuer Religion* (Oxford, 1956).
6. The relevant issues are interestingly treated in John Kekes' book, *A Justification of Rationality* (Albany, NY, 1976), 137–49.
7. For an interesting critique of cognitive relativism that is akin in spirit though different in orientation from that of the present section see Lenn E. Goodman "Six Dogmas of Relativism," in Marcello Dascal (ed.), *Cultural Relativism and Philosophy* (London, 1991), pp. 77–102.

8. To be sure, someone could convince me that my understanding of the implications of my standards is incomplete and lead me to an *internally* motivated revision of my rational proceedings, amending those standards from within with a view to greater systemic coherence.
9. *The Will to Power*, tr. by Walter Kaufmann and R.J. Hollingdale (New York, 1968), p. 481.
10. "Reste von christlicher Theologie innerhalb der philosophischen Problematik" (Martin Heidegger, *Sein and Zeit* (Leipzig, 1923), p. 230).
11. William James, "Pragmatism and Humanism," in *Writings*, ed. by John J. McDermott (New York, 1967), p. 450.
12. I am grateful to Charles Taliaferro for some constructive comments on an earlier version of this discussion.

CHAPTER 11

THE DEFICITS OF DECONSTRUCTIONISM

1. The Project of Deconstruction

Deconstructionism is a theory regarding the interpretation of texts. Initially projected with regard to *literary* texts, its more ambitious exponents soon expanded its application to texts in general – historical, philosophical, what have you. Its essential core involves two theses, first, that a text always allows many alternative interpretative constructions whose elaboration is the proper mission of the interpretative enterprise, and second, that all these various interpretations are more or less co-equal in merit – that none can be rejected without much ado as unsuitable, inappropriate, incompetent. The enterprise of text interpretation thus always confronts us with an irrevocable plethora of alternative and more or less equivalent possibilities.

At the core of the theory of deconstruction lies a view of textual plasticity – that as the enterprise of text interpretation proceeds, it brings to view an ever increasing range of viable alternative interpretations. On this basis, the partisans of deconstructionism condemn with the dismissive epithet of "textualism" the view that a given text has "a meaning" in such a stable and objective way as to favor one interpretation rather than another.

Insofar as deconstructionist theory represents a *doctrine* rather than a methodological *attitude* about text interpretation, it is a position based on a group of hermeneutical views or contentions which may be sketched roughly as follows: In the domain of text interpretation we face a situation of –

(1) *Omnitextuality*: Any proposed interpretation of a text must itself take the form of simply another text. In the hermeneutical sphere there is no way of exiting from the textual domain.

(2) *Plasticity*: Every text has multiple interpretations – it admits a plurality of diverse constructions.
(3) *Equivalency*: Every interpretation is as good (valid) as any other. These various constructions of a text (viz., interpretations) are all of equal or roughly equal merit: none is definitive, canonical, discriminatively appropriate – indeed none is substantially more cogent or tenable than the others.

It follows from these theses that in interpreting texts we always confront a plurality of (roughly co-meritorious) variants. Text interpretation admits of no rational validation or invalidation for one resolution over against another. It is always simply an exercise of free imagination: a project in which we can do no more than to explore interesting possibilities and cannot hope to validate a particular result as optimal in a cogent and stable way. Where issues of interpretation are concerned, we can only explore alternatives and cannot substantiate particular resolutions. Accordingly, we should never ask what a text *does* mean, but only what it *can or might* mean. In the realm of text interpretation there are no forced choices: it is an inherently indecisive enterprise – a fact that, happily, manages to "liberate us from the prison-house of language."[1]

Now there is little point in quarreling with premisses (1) and (2) of the preceding argumentation, seeing that the former is an obvious and evident truth, and that the latter is amply substantiated by historical evidence. And it follows from these two theses that any interpretation itself admits of variant interpretations. The problematic crux of the argument for a relativistic indifferentiation of text interpretation is thus premiss (3), with its assertion of merit equivalency. But is this premiss tenable? Is the hermeneutic realm indeed a free-for-all ruled by the idea that all interpretations are created equal? Is the textual interpreter indeed wandering within a hall of mirrors, wholly unable to implement the distinction between appearance and reality?

2. The Importance of Context: The Flaw in the Deconstruction's Ointment

The idea that merit-undifferentiated textuality is all there is in matters of interpretation totally overlooks a crucial fact. Interpretation – and the texts through which they are conveyed – are not created equal: some make sense, some only nonsense; some are ambiguous (have many plausible interpretations) others are more definite; some convey much information, others little; some state truths, some falsehoods. What the fallacy of indifferentist relativism of text interpretation overlooks to its own decisive detriment, is the crucial matter of *context*.

The first is that while it is indeed true that every text interpretation is itself a text, some nevertheless have a better systemic fit than others. They fit more smoothly and adequately

(1) into the wider context of text in general;
(2) into the larger context of the environing realm of thought and action.

And this consideration is crucial for the issue of interpretative adequacy.

The consideration that has to be urged against the hermeneutical egalitarianism of mainstream deconstructionism is that the textual realm is *not* closed and that there is a real world out there and that not every reality is virtual. Outside the textual realm there are observations and non-communicative processes and interactions with nontextual artifacts. There are not only tennis rule-books and tennis manuals but also tennis courts and players and games. Texts interconnect with reality through the mediation of intelligent agents. In sum, texts have a context. And this means that text interpretation is not a matter of free floating imagination – it is a matter of *scholarship*.

The process of *deconstruction* – of interpretatively dissolving any and every text into a proliferation of different merit-equivalent constructions is counterbalanced by the process of *reconstruction*. This pivotal hermeneutic process calls for putting texts into their larger contexts. After all, texts inevitably have a setting – historical, cultural, personal – on which their actual meaning is critically inpendent. And this contextual setting is far broader than other texts. In particular it comprises both processes (know-how) and products relating to known action outside the textual domain:

- noncommunicative practices in relation to certified production);
- communicative practices (usage);
- processes, procedures, methods, modus operandi in relation to modes of hiring;
- products of noncommunicative processes and practices;
- traditions of practices.

Text interpretation is an evidential exercise where one has to make the best possible use of the relevant data over a wide range of information. For it is clear that a wide variety of hermeneutical factors come into play with text interpretation:

1. What the text itself explicably affirms.
2. Other relevant discussions by the author bearing in the issues that the text addresses.
3. Biographical evidence regarding the author's education, contacts, relevant interactions with contemporaries, and the like.
4. Considerations of "intellectual history" regarding the state of knowledge and opinion in the author's place and time.
5. Philological data regarding the contemporary use of terms and expressions and the *modus operandi* of geometrical constructions.

All these factors bear on our understanding of texts – and bear differentially upon those variant constructions.

The crucial point, then, is that any text has an historical and cultural *context*. And this context constrains and delimits the viable interpretations that it is able to hear. The context at issue has three distinguishable levels:

- IMMEDIATE: Other parts of the same text.
- NEARBY: Other cognate discussions by the same author; other cognate discussions of the same genre or in the writings to which the author is responding by way of development or opposition.
- DISTANT OR PERIPHERAL: General aspects of the state of information and opinion of the time; general linguistic and philological considerations, etc.

Considerations at all these levels stand in the way of our equating the merits of all those different constructions and interpretations of a text by circumscribing the range of acceptable understandings that a text is able to bear.

And last but not least is the fact that texts are human artifacts produced by flesh-and-blood individuals along with innumerable other artifacts (buildings, utensils, etc.) in a collaborative endeavor to create a physical and communal environment where they and their fellows can flourish and thrive. And so we also come to yet another crucial hermeneutical factor, additional to the five listed above –

6. The setting of a nonverbal *modus operandi* within which texts take them place and play their role.

The upshot, then, is that the context of a text is not just textual – not something that can be played out solely and wholly in the textual domain. It also has another dimension in the setting of the wider praxis of which it is a part.

Neglect of this important factor leads to what is perhaps the most severe shortcoming of deconstructionism, betokening its deeply problematic commitment to the idea that the textual realm is self-sufficient and autonomous. Such a stance reflects the bias of academics committed to a logocentrism that sees the world in terms of discourse and forgets that it is not the case that everything is a matter of language through and through. Our texts and our use of words are, for the most part, no more than just one other instrumentality by which we function in a nontextual world – in this instance previously the social world of human interaction. To see texts and the libraries that warehouse them as self-sufficient is like contemplating the molehills without the mole.[2]

To be sure, all of the various interpretations of a text that are not totally bizarre have (by hypothesis) some sort of merit – there is *something* to be said for them. But to affirm this is not, of course, to say that the merit of all those different (nonabsurd) interpretations is thereby equal. To concede the prospect of a plurality of alternative nonabsurd interpretations is certainly *not* to say that any such interpretation is every bit of good as any other.

The idea that any old construal of a text – any bending or twisting of its message – is as good as any other is particularly dubious with any text that has a how-to aspect, whether this be small scale (recipes for baking bread, instructions for cleaning a rifle) or large scale (explanations for comprehending pre-Socratic ideas, prescriptions for successful salesmanship, guidelines to understanding Greek poetry). In such matters there is no anything goes plasticity; the simple fact is that some ways of interpreting that text and implementing the lessons of such an interpretation are materially better than others. The merit of deconstructionism lies in its stress on the importance of texts in humanistic studies and on the pluralism of interesting, discussable, and attention-worthy interpretations. But its defect lies in the idea that interpretations are created equal – that issues of quality and cogency are out of place in this domain.

A noticeable theory of text interpretation must be *normative*: it must be predicated on standards and criteria that provide for the evaluation of better and worse of sensible and foolish, of responsible and irresponsible. Sensible text interpretation is not a matter of anything-goes imaginative flights into the never-never world of free-floating fancy; it is tethered to the down-to-earth realities of the case imposed by rational standards of validity and appropriateness.[3] It is a matter of reconstruction rather than deconstruction.

3. Coherence as an Interpretative Standard

The crucial task of text interpretation is one of not merely *examining* possibilities but of *evaluating* them. Of pushing beyond the survey of *possible* interpretations to assess which of them are *plausible* and – going even beyond this – to decide which (if any) among them is *optimal*. But how to implement this project?

It is a profound error to see the textual sector is closed – to take the line that it is all a matter of texts "all the very through." Texts come into contact with contexts. The cardinal instrumentality of text interpretation is represented by the principle of *hermeneutical optimization* according to a standard of merit provided by the coherence of the proposed interpretation of a text with its overall context. Whatever interpretation best harmonizes with a text's overall context is *ipso facto* a superior interpretation which thereby has greater claims on our acceptance. In the light of such contextual considerations, texts are emphatically not created equal.

The most sensible approach to the existence of a variety of alternative text interpretation is what might be called the *coherence theory* of interpretation. The approach of this theory is predicated on two main theses:

1. The ultimate object of the interpretative enterprise is *optimization*. Its goal is not just to survey possible interpretations but to assess their respective merits and – above all – wherever possible to determine which one is the best.
2. The optimal interpretation of a text is that which can best achieve a *systemic unification* of the whole range of the hermeneutical factors previously enumerated (context, author data, philology, intellectual history and the rest). The determinative issue is that of the best overall fit, leaving the least overall residue of residual questions, problems, difficulties, loose ends, or the like.

Whatever can be said against a coherence theory of *truth*, a coherence theory of *interpretation* is eminently sensible. The fact that there are standards – that the situation we face with information-conveying texts is not a matter of an unfetteredly imaginative, anything-given, free-wheeling word-spinning – brings rationality on the scene once more. Text interpretation is a practice that can be more or less adequate in the light of the ultimate goal of *systematization*: of fitting texts into context in a way that realizes a systemic framework for understanding. Not only can a text have a *subtext* of implicit but inarticulated messages but it also – and more usually – has a *supertext* – a wider contextual environment within which its own message must be construed.

4. Against Textual Egalitarianism

It is thus necessary for sensible text interpretation to reject the mistaken idea that we can afford to forget the existence of an extra-textual world with which we humans interrelate on the basis of texts. The use of words is not something free-wheeling that stands disconnected from the verbal and behavioral environment in which they figure. The textual realm is not disconnected from the realm of human praxis. (Indeed even moving into and through that textual realm is a matter of praxis – producing and consuming texts is a matter of *doing* things.) The context of a text is set not only by other related texts but also by the artifacts that constitute its material environment and by the common elements of experience that we are ourselves declined to share with that text's author in virtue of the fact of sharing a common experiential framework in a shared human setting in a common world.

Jaques Derrida's critique in his *Speech and Phenomena*[4] that Western philosophical thought has privileged "voice" (authorial subjective thought) over "writing" (the resultant objective text) – process over product – involves a profound misconception. It overlooks the import of the hermeneutic circle: the lesson that in the endeavor to understand texts issues of process and product are coordinate: that neither can be subordinated to the other – let alone consigned to oblivion. To adopt the standpoint proposed by Derrida, and subordinate process ("voice") to product (the textual product in "writing") is to neglect the crucial lesson that what a text *is* depends on what it *does* and therefore on the context and its reception of its production.

The reason why interpretations are not created equal lies in circumstance of their contextual embedding in "voice"-related matters. Texts issue invitations to thought and thought issues in action. And actions (even intellectual action like understanding) can be more or less successful. Accordingly we can evaluate texts (representations) because they have a pragmatic dimension. Texts are instruments of communication – of conveying information and canalizing to action – even where the only "action" at issue is one of deliberation or discussion. And as such are elements of a wider teleological domain. Their adequate interpretation pivots on this fact.

What words mean is a matter of convention of subjective and decisional modus operandi of human linguistic and symbolic arrangements. But of course what can *appropriately* (warrentedly, correctly) be said once those controversial arrangements are in place is something that is itself no longer free-floating and decisional. Once we decide what "cat" and "mat" (etc.) mean, the question of whether one can appropriately assert "The cat is on the mat" is not an issue open for decisional resolution. In the lives of rational beings, words and actions are interconnected.

Medical texts influence the medications we take. Engineering texts influence the projects we construct. Literary texts influences the values we maintain and the lives we lead in their basis. Philosophical texts influence our priorities and the way we conduct our intellectual affairs. Texts have a bearing not only upon what we think but upon what we do: our actions and activities, our experiments and observations, our predictions and ventures at control. The textual world is *not* self-contained, it is inextricably interconnected with the realm of action, activity, and living. Even merely belletristic texts can guide people towards implementation in beliefs and values in ways that can be pointless and meretricious – or the reverse (to say nothing about matters of teaching and writing). Texts can be veridical or mendacious and which they would have or can be helpful or hurtful. They are not disconnected from life: they can be life-enhancing or life-degrading, can impel us to views that conduce to self-enhancement or to self-loathing.

The metaphysics of Derrida's *Of Grammatology*[5] – that organon of deconstructionism – pivots on an assault on the existence of abstract "meanings" that are somehow mysteriously present to the mind of interpreters: it is directed against the *logoi* of the Stoics as mediated through the subsequent idea of intellectual insight into essences, as contemplated in Husserl's *Wesenschau.*[6] But one need not be an aficionado of this latter-day version of neo-Platonic essentialism to uphold a theory of meaning that allows the operation of *standards* of interpretation. The idea that some interpretations of a text are better than others does not need the prop of a metaphysic of abstractly self-subsistent meanings. The limits to text interpretation are not set by the abstractions of a Platonic realm of intellectually accessible ideas, but by the operative realities that govern the relationship of those texts to the world in which we live and move. For the use of language does not – or need not – carry us into a realm of Platonic abstractions. What our words stand for – what we *mean* by "cat" or "mat" or "dog" is entirely a matter of human arrangements – of the "decision" of linguistic communities. But once these matters are fixed the question of whether and where and how frequently dogs and cats are to be encountered in nature – and in the proximity of one another and of mats – is something that only nature can resolve; the conventional arrangements of language-using communities has nothing further to do with it; the world's concrete realities now gain the upper hand.

Flat-out realism in communicative matters is an untenable exaggeration. We cannot write off the conventional, subjective, decisional elements in the production and processing of texts. But there are limits. In the study of communicating productive texts we will always reach a point where conventionality, decision, and subjectivity come to the end of their tether and where what can appropriately be said and thought depends on no *further* factors of this sort but hinges in "the world's objective facts."

In the domain of meaning and truth there is always some conventional element – an inevitable concession to idealism. But that its role is limited – that the appropriateness of thought and assertion in matters of communication will generally depend on factors from outside this domain – represents an unavoidable concession to realism.

The critical flaw of a deconstructionist relativism of texts and their interpretations lies in the fact that rational evaluation is possible in the sphere of text interpretations. Texts can be evaluated in terms of their contextual ramifications. It simply is not the case that everything in the textual realm is created equal – as deconstructionist relativism would have us believe.

5. Ideological Concluding Observations

The situation is not, however, entirely one-sided. There is something, after all, that can be said for the deconstructionist perspective. But interestingly enough, it falls wholly outside the sphere of strictly hermeneutical *interpretation*. For when we speak of "interpreting" a literary text, two distinctly different things can be at issue. It is crucial to distinguish between:

- *exegetical interpretation* – the endeavor to elucidate the meaning of a text in relation to the intentions of its producer *vis-à-vis* the intended audience then and there;
- *imaginative (re-) interpretation* – the re-casting or re-presentatioı of a text in an endeavor to evoke aesthetic responses and affective resonances in a current (present-day) recipient.

In doing actual (exegetical) interpretation we operate in the domain of scholarship. Here issues of context become centrally important because the pivotal question is "What did *the author* mean by the text?" The point of concern is with the original meaning and purport of the text. The issue of historical authenticity is paramount. But a text – and not only a text but any artifact that has an "esthetic" such as a painting or sculpture – can also be regarded abstractly, in a context-independent way. Here the issue is not one of producer centrality but one of consumer centrality and to issue is not "What *does* the text mean for its author?" but rather "What *can* the text mean for us?" Insofar as "interpretation" is at issue is not the hermeneutical interpretation of meaning explanation at all, but rather the sense of the term at issue when we speak of a performer's "interpretation" of a musical composition or the director's "interpretation" of a play. What we do here is not so much *interpret* the text as to creatively *reinterpret* in or endeavor to endow it with current relevancy and influence. Producing a play or a musical composition affords a paradigm example. Here we are (usually) not trying for historical authenticity but for the enlistment of interest. We are not addressing issues of scholarship but issues of edification or entertainment. Here authorship (and with it context) becomes of subsidiary importance – and imaginative creativity comes to the forefront. Where this sort of enterprise is at issue, the free-wheeling inventiveness envisioned by deconstructionisms has something to be said for it. But the hermeneutical commerce with texts geared to the enterprise of understanding their actual *meaning* is of course something quite different.

The crux, then, is the contrast between text interpretation as historically minded exegesis as against an imaginative de- (and all too often mis-) construction of texts freed from considerations of historical context – and thus with issues of scholarly exegesis in suspension. And behind this duality of approach looms a far-reaching quarrel in the area of educational approach and policy. The question is who owns the texts – by whose ground rules of textual "interpretation" is the game to be played. Are we to deal in scholarly exegesis or in imaginative edification. Are the text consumers at issue expected to bring scholarly industry to bear or are they being invited to indulge in imaginative flights of fancy. And, in particular, in conjuring with texts in the educational process, are we to advantage the philologically and historically well-informed, or are we to create a more level playing field of imaginative sensibility where anyone can play. Clearly, two very different "interpretative" enterprises are at issue.

The prospect of a *theoretical* reconciliation is thus at hand: "It's all a matter of how you understand interpretation. Different things can be at stake. Scholarly exegesis is one, deconstructively imaginative renovation is another." But behind this looms the deeper and more vexing question of primacy or priority: Which mode of "interpretation" deserves the pride of place in practice? Is education for learning or is it for consciousness raising and imaginative edification – is *understanding* or *appreciation* to enjoy priority? This deeply ideological battle is never far in the offing when the quarrels between post-modern deconstructionism and traditionalistic exegetes are being contested. And here – as with any other quarrel regarding the academic ideology – we can expect to encounter not peace but the sword.

NOTES

1. J. Hillis Miller in Harold Bloom, et al., (eds.), *Deconstruction and Criticism* (New York, 1979), p. 229.
2. This *a propos* of the quip that a book no more shows where its author is presently located in thought then a molehill shows where its maker is presently located in nature.
3. But who makes the rules? The answer is that they are not *made* but *given*. They are implicit and inherent in the broader context revolving both about the text itself and the purposive tradition within which the interpretation proceeds. An interpreter no more makes up the rules of the process on which he/she is engaged than a speaker makes up the meanings of the words he/she uses. (To be sure, both emerge from a tradition of human praxis and are not legislated from on high by a *deus ex machina*.)
4. Jaques Derrida, *Speech and Phenomena*, trans. by D. Allison (Evanston, 1973). Original: *Le Voix et le phénomène* (Paris, 1973).
5. Jacques Derrida, *Of Grammatology*, tr. by G. Spivak (Baltimore, 1976). Original: *De la Grammatologie* (Paris, 1967).
6. For a substantiation of this point of view see J. Claude Evans, *Strategies of Deconstruction* (Minneapolis, 1991).

CHAPTER 12

EXITS FROM PARADOX

1. STAGESETTING

The theory of paradox is an open invitation to unorthodoxy in logic and semantics. Many of the most far-reaching modern innovations in these domains have in fact developed through grappling with paradoxes. What is most interesting and significant about paradoxes, however, is not that they are paradoxical, but that they are so instructive.

Consider the following self-referential statement

(A) This very statement, *A* itself, is not the case;

or reformulated more simply:

(A) not-*A*.

This statement A is accordingly such that in virtue of its very definition:

A iff not-*A*.

Now suppose *A*. Then we can infer not-*A*. On the other hand, suppose not-*A*. Then we can infer *A*. Either way, we run straightaway into inconsistency. There is simply no way to make sense of *A*. The statement just is incoherent; we have no real alternative but to reject it as a meaningful assertion. So far no problem.

Again, consider the statement

(B) This very statement, *B* itself, materially)
implies its own denial,

or more compactly:

(B) *B* > not-*B* (where > represents the if then
relationship of "material" implication).

Then, by definition:

B iff $B >$ not-B.

Assume *B*. Then $B >$ not-*B* in virtue of *B*'s own definition. And now by *modus ponens* we obtain not-*B* – contrary to the assumption. But again assume not-*B*. Then not-($B >$ not B) in virtue of *B*'s definition. And this amounts to *B*, which once again contradicts the operative assumption. Thus, statement B is also incoherent. It too cannot be taken at face value: it can neither be asserted nor denied.

Such an analysis shows that the statements *A* and *B* of the preceding examples must both be rejected as meaningless. The logic of the situation is such that neither can be affirmed nor denied without plunging us into self-contradiction. Such statements are paradoxical to an extent that fatally compromises their own viability as meaningful units of communication. Can anything be salvaged from the wreckage?

2. A Version of the Liar

The paradoxical situation that confronts us becomes far less drastic when we shift from the case of outright affirmation and denial to claims of truth and falsity – say from (A) above to its cousin:

(C) *C* is not true.

For here, in this equally paradoxical-seeming case, various semantical expedients regarding the ways of truth and falsity can come into play to save us from outright self-contradiction.

Thus consider, for example, the (C)-analogous statement represented by the well-known Liar Paradox:

(L) This very statement, *L* itself, is false.

It is easy to see why one is tempted to refuse classing *L* as either true or false. Let us adopt $|P| = X$ as abbreviation for "The truth value of P is X," letting T and F represent the standard truth values of truth and falsity, respectively. Then L comes to $|L| = F$. Now suppose that L is true – that is, $|L| = T$. Then we can reason:

(1) $|L| = T$ by assumption

(2) not-($|L| = F$) from (1), assuming that the truth value specifying double-bar function is single-valued.

(3) L iff $|L| = F$ by the definition of L

(4) not-L from (2), (3)

(5) $|\text{not-}L| = T$ from (4) by the Tarski principle

(6) (5) conflicts with (1).

On the other hand, suppose that we class L as false: $|L| = F$. Then we can reason as follows:

(1) $| L | = F$ by assumption

(2) L iff $| L | = F$ by the definition of L

(3) L from (1), (2)

(4) $|L| = T$ from (3) by the Tarski principle

(5) (4) conflicts with (1)[1].

We have not, however, been brought to the end of our tether at this point. For the difficulty into which the Liar thesis (L) has plunged us is not absolute: it is a collision in which various other theses, apart from the Liar itself, also feature as participants whose contributory role enroute to self-contradiction is actually indispensable. And harmony can, in principle, be restored by laying the blame of paradox not at the Liar's door, but elsewhere.

Note in particular that in the preceding deliberations all of the following principles have come into play:

(0) *Meaningfulness*: Even paradoxical claims like L belong to the realm of meaningful statements and thus fall into the range of the propositional variables P, Q, etc. by whose means we articulate logical and semantical principles.

(I) *Single valuedness*: If $|P| = X$, then not also $|P| = Z$ for any $Z \neq X$

(II) *Bivalence*: Either $|P| = T$ or $|P| = F$

(III) *Tertium non datur*: Never $|P| = Z$ for $Z \neq T, F$ [Note, given (I), this is equivalent to (II) – although not so in (I)'s absence.]

(IV) *Tarski–1*: If P, then $|P| = T$

(V) *Tarski–2:* If $|P| = T$, then P

(VI) *Negation*: $|P| = F$ iff $|\text{not-}P| = T$

The preceding argumentation shows that once we adopt this entire collection of plausible semantical principle, (L) plunges us into an outright contradiction.[2] But it shows no more than that. And in particular it does not establish L's absolute untenability, independently of other considerations. There is no need to lay the burden of inconsistency at L's door.

3. A Choice of Exits

In any aporetic situation of this general sort we have choices: rational conflicts within a group of contentions always admit of diverse resolutions, depending on which parties to the conflict are made to yield to the others. And so in the case at hand we have various alternatives, preeminently including the following three:

1. *Paradox exclusion*: to eject paradoxical statements such as L from the domain of the meaningful, and accordingly to characterize such paradoxical statements as being non-assertions. *Tactic*: to reject principle (0) and treat claims of the Liar type in the same way as the statements (A) and (B) considered at the start of the paper.
2. *Truth-value gaps*: to accept the idea that some statements (viz., the paradoxical ones) are *neither true nor false*. Paradoxical statements as now seen as semantically INDETERMINATE in lacking a (classical) truth value. *Tactic*: to reject the *tertium non datur* principle of thesis (III). This tactic of abandoning principle (III) does *not* require us to also reject the bivalence principle (II) and introduce a new, "indeterminate" truth value (say I for "indeterminacy"). For paradoxical statements can be characterized as such not by their having an eccentric, non-classical truth value, but rather by altogether *lacking* a truth value. (From the no-value point of view, their truth *status* is then defined by just exactly their lack of a truth *value*.)
3. *Tarski abandonment*: Abandon the Tarski principle. Tactic: give up IV/V.

It deserves note that the attempt to eliminate the difficulty at issue with propositions akin to the Liar by introducing I as an additional ("indeterminate") truth value will not of itself extricate us from difficulty. For paradox now reenters through the back door. Thus consider the statement,

(K) This statement, K itself, is not indeterminate but false,

or equivalently:

(K) $|K| \neq I$ and $|K| = F$.

We are now trapped. For consider:

A. Assume K is true, that is, $|K| = T$. Then K's second conjunct fails. So K is false, contrary to hypothesis.

B. Assume K is false, that is, $|K| = F$. Then also $|K| \neq I$. So K is true because both its conjuncts obtain. This too is contrary to hypothesis.

C. Assume K is indeterminate, that is, $|K| = I$. Then K's first conjunct fails. So K is false, again contrary to hypothesis.

The lesson is that if we wish to take the truth-value gap approach to paradox resolution then we had best interpret truth-value indeterminacy in the just indicated way, as a nonstandard truth *status* vis-à-vis the classical truth values rather than as a new additional truth *value*.

A variant approach is also possible, however – one that in effect accepts paradoxical propositions at face value, taking their paradoxicality to consist exactly in their being truth-status anomalies in virtue of being *both true and false*. This yields yet another alternative approach:

4. *Truth value overlaps*: to accept the idea that some statements (namely, paradoxical ones) are AMBIVALENT in being *both* true *and* false. Such paradoxical statements are now seen as truthwise over-determined by their possession of both of the (classical) truth values – that is, by their anomalous feature of counting as both true and false.[3] *Tactic*: to reject the principle of single-valuedness (I) and in consequence to modify principle (V) to the variant version:

$$(V') \text{ if } |P| = T \text{ and not also } |P| = F, \text{ then } P.$$

A small but important conceptual gap now opens up between truth and assertability because only with nonparadoxical statements is the move from truth to assertability automatically authorized.

Such an approach would enables us to take the Liar thesis entirely at face value, one part of it establishing the truth of the statement at issue, and the other part its falsehood – and thereby its paradoxiality. But of course the shift from (V) to (V′) indicates that we will not be prepared to see the Liar thesis as *assertable* – its conceded (ambivalent) truth to the contrary notwithstanding. The shift in question constrains us to complicate the linkage of truth to assertability.

In contemplating truth-value *overlaps* we wish to be able to treat paradoxical propositions as non-assertable, exactly as in the more orthodox approach of a recourse to truth-value *gaps*. To be sure, this requires some systemic readjustments. "The connection between truth and assertion" a contemporary logician informs us, "resides in the fact that the assertion of a sentence is its assertion *as true*."[4] But even if this is so, there is no automatic impediment to going on to add "as true *inter alia*," or perhaps even "as true *and not false*." Nothing in the abstract logic of the situation – or even in our informal understanding of the issues – blocks the prospect of reconstructing the well-known Tarski principle into the form:

$$P \text{ iff } | P | = T \text{ but not } | P | = F.$$

Abstract logico-semantical theory as such is flexible enough to accommodate such a step – and our informal understanding of the issues is surely not so precise as to constrain a ruling one way or the other. (After all, from the accustomed classical point of view that addendum is harmlessly redundant.)

In modifying (V) to (V′) we thus effectively adopt the idea of *ambivalent truths* – that is, propositions that are true alright, but not *definitely* true – that is, true to the exclusion of also being false (thereby manifesting themselves as not being "eccentric," "paradoxical," or the like). And we will accordingly have to complicate the link between truth and assertability, seeing only unambivalent truths as assertible, and not truths in general and as such.

Now however eccentric our propositional *semantics* may have to become under the pressures imposed by paradox accommodation (that is, however nonstandard may be the principles we employ in the business of truth-value assignment), we shall – for present purposes at least – want to keep to principles of propositional *logic* (that is, of consequence and incompatibility) as orthodox and classical as is possible in the circumstances. Thus we wish is to avoid the *logical* inconsistency of ever having to give assent on our own account to both P and not-P. And a semantics of paradox that sees paradoxical statements as truth-value anomalies that are both true and false certainly does not constrain this result. Not, that is, unless we retain both of the classic semantical principles:

$$\text{If } |P| = T, \text{ then } P$$

and

$$\text{If } |P| = F, \text{ then not-}P.$$

But of course these principles leave the scene once we modify Tarski–1 – that is (V) – to obtain its revised version (V′) as indicated above. This bit of *semantical* complication permits us to retain *logical* orthodoxy.

It is instructive to note that the abandonment of Tarski–1 – principle (V) – in the context of paradoxical theses such as L is pretty much unavoidable in any case – even for the partisans of truth-value gaps (rather than overlaps). For by (V) we would have it that

$$\text{If } |L| = T, \text{ then } L$$

or equivalently, by contraposition:

$$\text{If not-}L, \text{ then not-}(|L| = T).$$

Now by the very definition of L we have

$$L \text{ iff not-}(|L| = T).$$

Seeing that these theses together entail "If not-*L*, then *L*," we have the result that *L* must obtain in any event. And given the paradoxical nature of *L*, this consequence of having to maintain *L* is clearly unacceptable.

Thus while our plunge into unorthodox semantics need not have any deforming consequences for our logic as such (our principles of consequence and incompatibility), it does, nevertheless, require some readjustment at the level of the bridging principles that link our logic with our semantics. The most of these prominent is perhaps:

> If the premisses P_1 , P_2 ,..., P_n logically entail the conclusion Q – symbolically if P_1 , P_2 ,..., P_n => Q, then it cannot be *that the premises are all true and the conclusion false*.

With truth-value overlaps, this principle becomes unavailable. For clearly *P* entails *P* – so that *P*=>*P*. But if *P* is paradoxical, then it can happen that the premise of this valid inference (viz., *P*) is true and nevertheless the conclusion (viz., *P* once again) is false. So we need to modify the preceding principle to read in its final clause that "then it cannot be *that the premisses are all true and the conclusion is not true*."

And various other logico-semantical bridging principles will need some corresponding adjustment. Take, for example, the principle that what entails a falsehood cannot be true:

If A=>B and $|B| = F$, then not-($|A| = T$).

Clearly if *P* happens to be ambivalent, then we run into trouble in the particular case: $A = P$ and $B = P$. We have to make the necessary readjustments to reckon with the fact that when ambivalence comes on the scene, then falsehood ($| P | = F$) is no longer tantamount to untruth (not-($| P | = T$). But all this calls for a no more than routine readjustment of classical principles.

It deserves comment that there is a certain parallelism between the truth-value gap and the truth-value overlap approach – between locating the anomaly of paradoxical statements in a truth-value underdetermination and a truth-value overdetermination. Both approaches, it could be said, treat the same basic phenomenon in a parallel, through reciprocally inverted way. Nevertheless from the standpoint of logico-semantical *principles*, the two approaches represent very different processes and priorities. For while both alike call for a modification of the Tarski principle (V), the truth-value overlap approach is able to retain it in the alternate version (V′). And while truth-value gaps call for abandoning (I) while retaining (II), truth-value overlaps call for abandoning (II) while retaining (I). Parallelism at the level of truth-value assignments nevertheless leaves room for substantial divergence at the level of the operative semantical *principles*.

It is, of course, in theory possible to *combine* the two approaches, adopting a complex semantics that embodies *both* truth-value gaps *and* truth-value overlaps. But, seeing that either approach alone provides machinery sufficient to accommodate the paradoxes, there seems no particular point in going to such lengths. It seems sensible to minimize one's recourse to unorthodoxy relative to the overall needs of the situation.

4. Retrospective Deliberations

The main lesson of these deliberations is straightforward. It is that paradox dismissal and truth-value-gap acceptance are not the only viable responses to semantical paradoxes such as the Liar. Truth-value *overlap* offers an additional option which, despite its air of unorthodoxy, enjoys certain substantial advantages over its rivals – in particular the not insignificant advantage of making it possible to treat the semantical paradoxes in an intuitively natural way.

Moreover, a semantics of truth-value overlaps also affords a mechanism for dealing with *borderline* statements. For example, claims about color characterization in the "gray area" between (say) red and orange – a region of puzzlement where one could equally well (or ill) say both "This patch is orange" and "This patch is red." Such a situation could be handled not by classing these claims as truth value *underdetermined* (i.e., neither true nor false), but rather by seeing them as truth value *overdetermined* (i.e., both truth and false). The fuzzy "gray area" between fully determinate situations would thus be viewed not in terms of an absence of a (classical) truth value but in terms of the absence of a *definite* (classical) truth value – that is, of one to the exclusion of the other.

This point of view suggests that one can also deploy truth-value overlaps in much the same way as truth-value gaps to deal with other sorts of paradoxes. Consider the Sorites Paradox in the particular version represented by the following aporetic cluster:

(1) Patch A_0 looks red

(2) Patches A_i , A_{i+1} are pairwise color-indistinguishable for i = 0, 1,..., n - 1 (i.e., members of these pairs always have the same apparent color)

(3) Patch A_n does not look red

(4) Whenever A_i and A_j are color-indistinguishable, A_j and A_k are, then A_i and A_k are color-indistinguishable.

Here (1)–(3) appear to be "facts of life" and (4) seems eminently plausible. But the whole group is logically inconsistent, so something has to go. Seeing that (1)–(3) are phenomenological reports which, as such, seem totally unproblematic, our best available strategy here is to see (4) not as straightforwardly true and assertable but rather as ambivalent – that is, as involving an admixture of truth *and* falsity, the latter because falsehood emerges with a multiple application of (4) in the course of a single piece of reasoning. Accordingly, (4) would provide a contrast with the Liar statement in presenting an example of an ambivalent statement whose form is not particular but universal.

The truth-value overlap approach thus leads us to envision a paradoxical domain of ambivalent truths – statements to which we can assign the truth value T all right, but not in a way that also denies them the truth value F. However, in the circumstances here envisioned such statements now cannot be asserted categorically, seeing that

$$\text{If } |P| = T, \text{ then } P$$

fails us and we have instead is merely:

$$\text{If } |P| = T \text{ and not also } |P| = F, \text{ then } P.$$

The path from truth to assertability is now complicated by the prospect of *ambivalent* truth (in just the way contemplated above). And of course only unqualifiedly assertible statements are available premisses of categorical inference – though, to be sure, ambivalent truths can (like actual falsehoods) be assumed or supposed for the purpose of hypothetical inference, *reductio* argumentation, or the like.

As these deliberations indicate, the idea of ambivalent truth enriches our logico-semantic arsenal in a way that is interesting and useful – albeit rather unorthodox. In particular, it yields the advantage of making the treatment of paradoxical statements at once natural and rationally manageable.

In a book on *The Logic of Inconsistency* written some fifteen years ago,[5] Robert Brandom and the author considered the idea of *inconsistent possible worlds* – possible worlds characterized by means of theses that are both true and false. And we suggested that the semantical paradoxes could be seen as amenable to treatment in these unorthodox terms. The present deliberations further implement this idea. In so doing they provide a vehicle for the rather striking prospect that this world of ours – in which, after all, paradoxes like that of the Liar and the Heap do actually arise – may in the end be most advantageously be seen as actually constituting such an inconsistent world. But note that adopting the Tarski principle only in the revised form (V′) means that our own discourse about such an inconsistent world, with its variegated proliferation of truth claims, should *and can* remain perfectly consistent. The semantical anomalies of a world in respect to its truths and falsehoods need not now drive our own discourse about it into logical incoherence.

But of course caution is in order. Inconsistency may not be an all-destructive H-bomb that levels everything in sight – there are ways to contain the damage and salvage materials from the wreckage. But it is unquestionably a negativity that we would certainly want to minimize. To repeat the strictures of Chapter 4 above, the tolerance of inconsistency has to be seen as a desperate expedient to be adopted only if and when all the available alternatives look worse. However, the logico-semantical paradoxes may well represent an uncomfortable situation of just this sort. Yet even so, reason as such has nothing to fear. Its job is to come to enable us to come to terms with the realities, and it can accomplish this even where those realities happen to be inconvenient.[6]

NOTES

1. Alternative route:
   (4) $|\text{not-}L| = T$ from (1)
   (5) not-L from (4) by the Tarski principle
   (6) (5) contradicts (3)
2. Specifically, principles 0 and II (or III) are involved in setting up the dilemma; I, IV and VI are involved in the first dilemmatic argument; and I, IV (or alternatively V, VI) are involved in the second dilemmatic argument.
3. Given the logic of the paradox-engendering argumentation, we must either reject one of the pair 0, II (or III) or else reject one of the trio: I, IV, VI, and I, IV, and V, VI. This offers just the following four maximally conservative

possibilities:

reject 0:	alternative 1 (paradox exclusion)
reject II (or III):	alternative 2 (truth-value gaps)
reject IV and either V or VI:	alternative 3 (Tarski-abandonment)
reject I and either V or VI:	alternative 4 (truth-value overlaps)

Since we here wish to be conservative as regards logic, and avoid espousing a non-standard mode of negation, we here treat (VI) as non-negotiable.

4. Michael Dummett, *Frege: Philosophy of Language* (London, 1973), p. 460.
5. Nicholas Rescher and Robert Brandom, *The Logic of Inconsistency* (Oxford, 1980).
6. The stimulus for this paper was provided by some discussions of cognate issues with my Pittsburgh colleague, Prof. Jamie Tappenden.

NAME INDEX

Episteme

A SERIES IN THE FOUNDATIONAL, METHODOLOGICAL, PHILOSOPHICAL, PSYCHOLOGICAL, SOCIOLOGICAL, AND POLITICAL ASPECTS OF THE SCIENCES, PURE AND APPLIED

1. W.E. Hartnett (ed.): *Foundations of Coding Theory.* 1974 ISBN 90-277-0536-4
2. J.M. Dunn and G. Epstein (eds.): *Modern Uses of Multiple-Valued Logic.* 1977 ISBN 90-277-0747-2
3. W.E. Hartnett (ed.): *Systems: Approaches, Theories, Applications.* 1977 ISBN 90-277-0822-3
4. W. Krajewski: *Correspondence Principle and Growth of Science.* 1977 ISBN 90-277-0770-7
5. J.L. Lopes and M. Paty (eds.): *Quantum Mechanics, a Half Century Later.* 1977 ISBN 90-277-0784-7
6. H. Margenau: *Physics and Philosophy.* Selected Essays. 1978 ISBN 90-277-0901-7
7. R. Torretti: *Philosophy of Geometry from Riemann to Poincaré.* 1978 ISBN Hb 90-277-0920-3 / Pb 90-277-1837-7
8. M. Ruse: *Sociobiology: Sense or Nonsense?* 1979; 2nd ed. 1985 ISBN Hb 90-277-1797-4 / Pb 90-277-1798-2
9. M. Bunge: *Scientific Materialism.* 1981 ISBN 90-277-1304-9
10. S. Restivo: *The Social Relations of Physics, Mysticism, and Mathematics.* Studies in Social Structure, Interests, and Ideas. 1983 ISBN Hb 90-277-1536-X / Pb (1985) 90-277-2084-3
11. J. Agassi: *Technology.* Philosophical and Social Aspects. 1985 ISBN Hb 90-277-2044-4 / Pb 90-277-2045-2
12. R. Tuomela: *Science, Action, and Reality.* 1985 ISBN 90-277-2098-3
13. N. Rescher: *Forbidden Knowledge* and Other Essays on the Philosophy of Cognition. 1987 ISBN 90-277-2410-5
14. N.J. Moutafakis: *The Logics of Preference.* A Study of Prohairetic Logics in Twentieth Century Philosophy. 1987 ISBN 90-277-2591-8
15. N. Laor and J. Agassi: *Diagnosis: Philosophical and Medical Perspectives.* 1990 ISBN 0-7923-0845-X
16. F.P. Ramsey: *On Truth.* Original Manuscript Materials (1927–1929) from the Ramsey Collection at the University of Pittsburgh, edited by N. Rescher and U. Majer. 1991 ISBN 0-7923-0857-3
17. H.A. Shenkin: *Medical Ethics.* Evolution, Rights and the Physician. 1991 ISBN 0-7923-1031-4
18. E. Agazzi (ed.): *The Problem of Reductionism in Science.* 1991 ISBN 0-7923-1406-9

KLUWER ACADEMIC PUBLISHERS – DORDRECHT / BOSTON / LONDON

Episteme

19. D. Blitz: *Emergent Evolution.* Qualitative Novelty and the Levels of Reality. 1992 ISBN 0-7923-1658-4
20. S. Restivo: *Mathematics in Society and History.* Sociological Inquiries. 1993 ISBN 0-7923-1765-3
21. N. Rescher: *Satisfying Reason.* Studies in the Theory of Knowledge. 1995 ISBN 0-7923-0000-0

KLUWER ACADEMIC PUBLISHERS – DORDRECHT / BOSTON / LONDON